In Canadian Service Aircraft

Handley Page
Halifax

Photo Opposite: Halifax B.MK.III at dispersal, 425 Squadron, autumn 1944.
(CF PL40183)

In Canadian Service Aircraft

Handley Page
Halifax

Anthony L. Stachiw
Andrew Tattersall

Vanwell Publishing Limited
St. Catharines, Ontario

Copyright © 2005   Anthony L. Stachiw and Andrew Tattersall

All rights reserved. No part of this book may be reproduced or used in any form or by any means, electronic or mechanical, including photocopying, recording, or in any information storage and retrieval system, without permission in writing from the publisher.

Vanwell Publishing acknowledges the financial support of the Government of Canada through the Book Publishing Industry Development Program for our publishing activities.

Vanwell Publishing Limited
1 Northrup Crescent
P.O. Box 2131
St. Catharines, Ontario L2R 7S2
sales@vanwell.com
tel: 905-937-3100
fax: 905-937-1760
Printed in Canada

Cover and Book Design by Renée Giguère
Cover Illustration by Stephen Otvos

Library and Archives Canada Cataloguing in Publication

Stachiw, A. L. (Anthony L.), 1940-
    Handley Page Halifax / Anthony L. Stachiw, Andrew Tattersall.

(In Canadian service. Aircraft ; 3)
Includes bibliographical references.
ISBN 1-55125-085-3

    1. Halifax (Bomber) 2. Bombers–Canada. 3. Canada. Royal Canadian Air Force–History. I. Tattersall, Andrew, 1971- II. Title. III. Series.

UG1242.B6S678 2005   623.74'63   C2005-903897-7

CONTENTS

CHAPTER ONE
Origins of the Design 7

CHAPTER TWO
The Halifax in Canadian Service 21

CHAPTER THREE
Aircraft Description and Drawings 27

CHAPTER FOUR
Halifax Squadrons and Units 59

CHAPTER FIVE
Halifax Colour Schemes and Markings 107

CHAPTER SIX
Aircraft Armament Configurations and Ordnance 115

CHAPTER SEVEN
RCAF Memorial Museum Halifax Restoration 120

CHAPTER EIGHT
Modelling the Handley Page Halifax 124

ACKNOWLEDGEMENTS

There are several people whose assistance has been instrumental in the writing of this book. I would, once again, like to thank John Griffin, the noted authority on Canadian military aircraft, and author of several books on the subject, for his encouragement and support. As well, John Turanchik has, over the years, provided articles from magazines whenever he spotted material relevant to the Halifax aircraft. A special mention is owed to Bill Scobie for his photographs of the models of the Halifax, and to John Bradley for information provided from his database on aircraft flown by RCAF Squadrons. Lt Col (Ret'd) Chris Colton, Executive Director of the RCAF Memorial Museum at Trenton graciously provided information on the RCAF Memorial Museum and the restoration of Halifax NA337.

Thanks also to the staff of the Canadian Forces Joint Imagery Centre, A Squadron, Department of National Defence; in particular, Janet Lacroix and WO Steve Sauve provided assistance and direction in the procurement of the photographs. The approval for their use was expedited by Simmy Chauhan of the Department of National Defence, Directorate of Intellectual Property.

Finally, thanks to Simon Kooter of Vanwell Publishing for providing the opportunity to present this book, and to Angela Dobler, the editor, for her expertise in its preparation.

CHAPTER ONE

Origins of the Design

Personnel watch the takeoff at dusk of a Handley Page Halifax bomber for a raid on enemy territory. (Canadian Forces [CF], PL-116971)

Handley Page Limited, the first aeronautical engineering company registered in the United Kingdom, was formed by Frederick Handley Page on 17 June 1909 as a private limited company. Its early projects were aircraft constructed to the requirements of other inventors. Later the company began to manufacture aircraft using its own designs. The company set up facilities at Woolwich, Fambridge and Barking Creek before establishing works at Cricklewood, North London and Radlett Aerodrome in Hertfordshire.

When the First World War began on 4 August 1914, Handley Page offered his factory and resources to both the War Office and Admiralty, but the War Office declined his services because he had failed to provide BE.2a aircraft on time. However he was called to a meeting with the Director of the Air Department, Admiralty, to discuss design plans for bombing and coastal patrol aeroplanes. The Admiralty was convinced of the need for a long-range heavy patrol bomber capable of attacking the German High Seas Fleet at its base at Kiel, as well as the growing number of Zeppelin airship sheds being constructed along the Friesian coast.

Handley Page, with his designer George R. Volkert, set out to design a much larger twin engine biplane, designated Type O/100. By careful attention to detail they achieved their design target weight, in an aeroplane of 100 ft wingspan which could meet the Admiralty specification. The structural design was established by February 1915. The Admiralty ordered four prototypes. The order was increased to twelve and then to a total of forty aircraft. By the end of November, the first prototype was ready for final assembly. The engine selected was the Rolls Royce 250 hp Eagle, which was further developed to produce 320 hp, thus improving the rate of climb and ceiling. First deliveries to the Royal Naval Air Training Squadron at Manston took place in October 1916. No. 7 Squadron was deployed to France in April of 1917 with six of these aircraft.

An improved version, the O/400, with smaller engine nacelles, powered by the Rolls Royce Eagle VIII was flown in September 1917. This variant had an increased fuel capacity and was capable of carrying larger bomb loads. When it was ordered into large scale production a new factory had to be built to meet this requirement, and component manufacture was subcontracted.

Handley Page then developed a long-range heavy night bomber capable of attacking strategic targets as far away as Berlin. Three prototypes of this V/1500, powered by four Rolls Royce Eagle VIII engines, were ordered. The production would take place at the Beardmore and Harland & Wolff factories, while design and development remained with Handley Page. Components manufactured at Harland & Wolff were shipped to the Handley Page facility at Cricklewood where the aircraft was built and flown on 18 May 1918. The first three V/1500s were operational by 9 November, but with the signing of the Armistice on 11 November, the operation to bomb Berlin was cancelled.

In December 1919 the prototype W.8 made its first flight, and in 1923 the W.8d military variant entered RAF service as the Hyderabad bomber. An enlarged three-engine derivative, the W.9, was offered in both civil and military versions. Only the civilian variant was ordered which served with Imperial Airways. That company had been formed by the amalgamation of Handley Page Transport Ltd. with its three British competitors to take advantage of a Treasury Board subsidy. Redesignated W.9a when converted to Bristol Jupiter engines, the aircraft gave excellent service. The airline also operated four W.10 twin engine airliners which were civilianized W.8d Hyderabad bombers. The Hyderabad bombers, when later converted to the Bristol Jupiter engines, emerged as the Hinaidi bomber or Clive troop carrier.

However, after 1924 the Air Ministry began to enforce a condition that all new contracts incorporate all-metal construction. Nevertheless, Handley Page designed and produced the HP.50 Heyford bomber, still using the biplane spaced-frame construction method. These aircraft remained in service until the late 1930s. In the company's first venture into the new stressed skin monoplane era, Handley Page produced an order for 100 HP.54 Harrow bomber-transports for the initial stage of the Royal Air Force expansion scheme. They also produced the much more modern HP.52 Hampden medium bomber.

In 1932 the Geneva Disarmament Conference opened, promoted by the League of Nations. Since the major world powers had earlier failed to agree on total disarmament, this conference was an attempt to agree on armament limitations. An attempt to impose a weight limitation on bomber aircraft was discussed over a considerable period of time, and finally rejected out of hand by a number of countries.

As the threat of war in Europe increased, Britain began a large-scale rearmament program. Although there was still talk of disarmament, events on the international stage were raising alarms. Germany and Italy and the other major powers had begun to rearm, and both countries, as well as the USSR, were already testing their weapon systems in the civil war in Spain. Although there was some suggestion of limiting the number of bombers in service, the Soviets already had four-engine bombers in service and other countries including the United States had similar designs under development.

As part of the modernization and expansion of the Royal Air Force Bomber Command, the Air Ministry issued Specification B.1/35, calling for a twin-engine replacement for the Armstrong Whitworth Whitley, Handley Page Hampden and Vickers Wellington medium bombers, none of which had yet entered service. The specification called for a bomb load of 2,000 lbs to be carried at a cruising speed of 195 mph at 15,000 ft over a range of 1500 statute miles. The wing span was limited to 100 ft so the aircraft could be accommodated in the standard RAF hangar design. The aircraft was to be powered by two British engines of 1,000 hp.

Handley Page Limited proposed their HP.55 design to that specification, powered by either two Bristol Hercules ISM or two Rolls Royce Merlin XX engines. The

CHAPTER ONE

The Vickers Wellington medium bomber was the mainstay of the RAF's Bomber Command night bombing campaign, along with the Handley Page Hampden and Armstrong Whitworth Whitley, until they were superseded by the four-engine heavy bombers. *(CF PL-15382)*

design featured a low mounted wing with a swept back leading edge, with the main undercarriage retracting into the rear of the engine nacelle. The twin fin and rudder layout and fuselage was similar to the HP.54 Harrow. Because the Department of Technical Development requirement was for as small an aircraft as possible, the bomb load was split between the fuselage bomb bay and bomb cells in the two-spar wing. Although a mockup was built with the serial number K8179, the proposal was rejected by the Air Ministry.

The Air Ministry later issued Specification P.13/36, dated 8 September 1936 and issued in November of that year, calling for a 45,000-lb all-metal, mid wing, cantilever monoplane medium bomber powered by two 24-cylinder Rolls Royce Vulture X-type engines developing 1,760 hp. The aircraft was to carry a bomb load of 8,000 lbs and have power operated nose and tail turrets. The design requirement set out in considerable detail the high standard of achievement expected. Handley Page designer George Volkert came to the conclusion that an aircraft designed to Specification P.13/36 would actually fulfill the requirement of Specification B.1/35, and he pointed this out to the Air Ministry in a letter dated 22 September 1936.

Proposals were submitted by A.V. Roe Co. Ltd (Avro), Handley Page Ltd. and Hawker Aircraft Ltd. Although the Vulture engine was still in the development stage, prototypes of the Avro and Handley Page proposals using this engine were ordered on 30 April 1937. The serial numbers L7244 and L7245 were issued to the two prototypes of the Handley Page HP.56 and a mockup was built.

As initial design work progressed, changes were made to the original design specification. The dive

The Handley Page HP.56 design proposal to meet Specification P.13/36 was redesigned with four Rolls Royce Merlin engines, to become the HP.57 Halifax. *(Andrew Tattersall)*

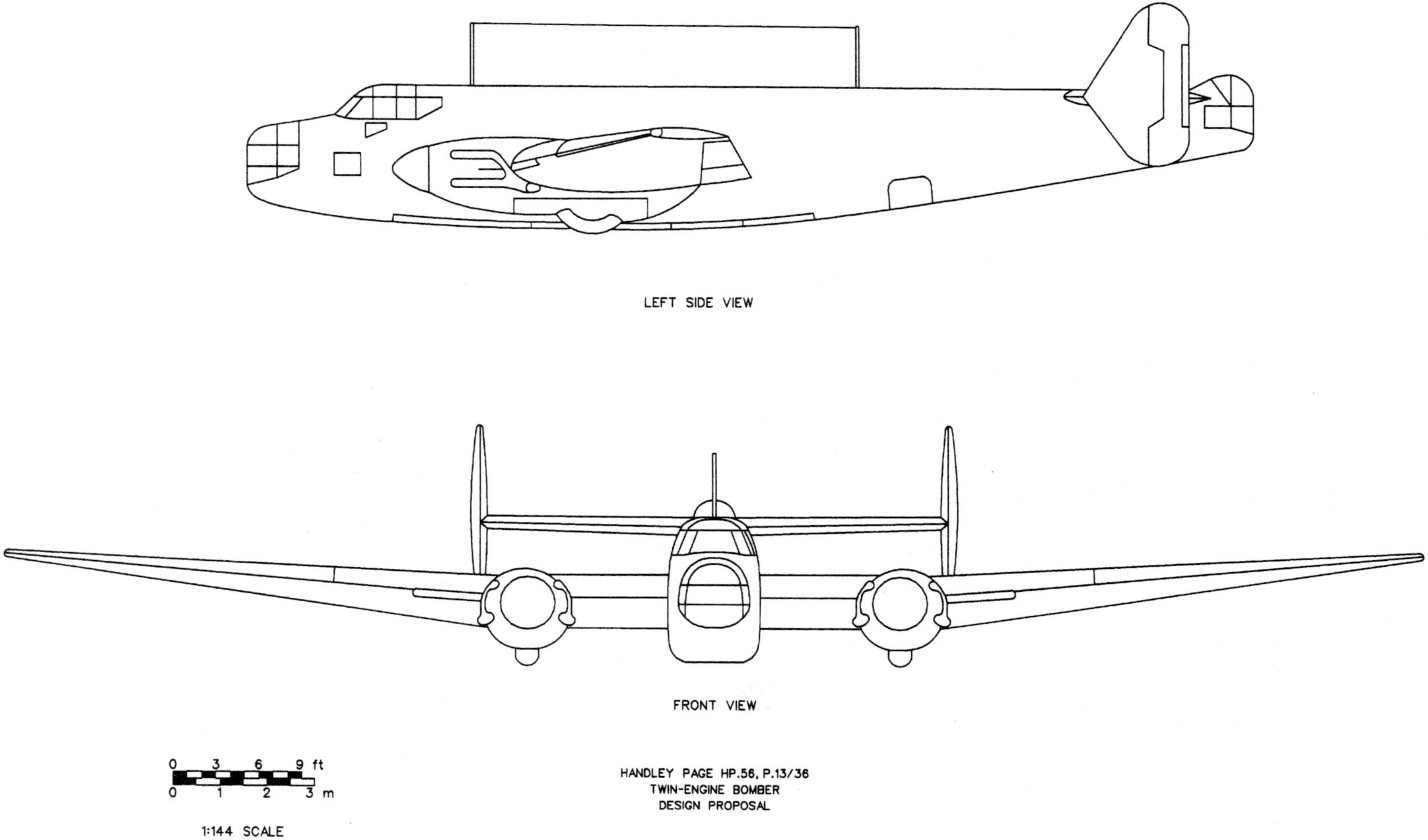

LEFT SIDE VIEW

FRONT VIEW

HANDLEY PAGE HP.56, P.13/36
TWIN-ENGINE BOMBER
DESIGN PROPOSAL

1:144 SCALE

TOP VIEW

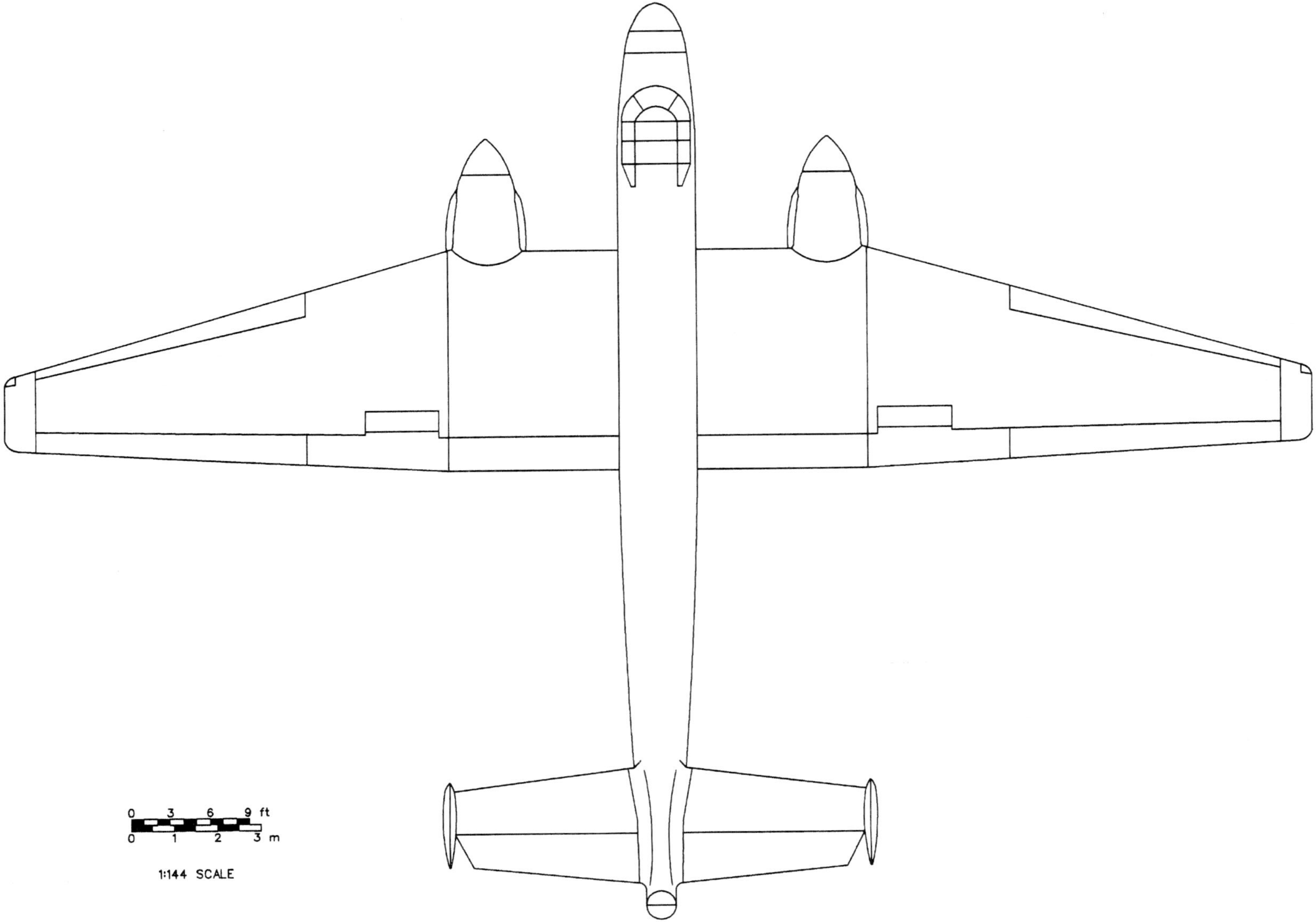

1:144 SCALE

Handley Page Halifax

bombing angle requirement was reduced to 25 degrees, and the torpedo carrying and message hook requirement were deleted. The maximum number of 1,000-lb bombs to be carried was reduced to 15 from 16. The resulting shortening of the bomb bay made room for an area behind it to be reserved for a ventral gun turret—the later H2S radar installation—or the exit hatch for paratroop drops or other special operations requirements.

The company accepted the risk of specifying the Rolls Royce Vulture engine, based on the reputation of Rolls Royce and their success with the Merlin V-12 engine series. But at least three years of development would be necessary to assure its reliability for operational service. However, serious problems arose with development of the Vulture engines, which had twenty-four cylinders, four banks of six cylinders arranged in an X configuration around a single crankshaft. Although work had begun on the first of the two prototypes, the HP.56 was shelved when it was realized that these engines would not be available in sufficient quantities to satisfy the demand.

Handley Page was asked to investigate various four-engine layouts: the Taurus, Kestrel, Dagger and Pegasus engines. The Bristol Hercules was not considered since it had been earmarked for the Short Stirling bomber. None of these powerplants were capable of overcoming the increased drag and weight imposed by the four-engine layout. Reginald Stafford, in charge of aerodynamics and performance at Handley Page, concluded that, when using four up-rated Rolls Royce Merlin engines, the takeoff performance was improved to the extent that the requirement for a catapult assisted takeoff was unnecessary. Finally, the dive bombing requirement was eliminated. On 3 September 1937 the Air Ministry ordered two prototypes powered by four Rolls Royce Merlin V-12 type engines under a different contract, using the serial numbers which had been allocated to the HP.56 design. The higher powered Merlin engines never materialized, these engines being reserved for fighter designs.

Avro proceeded with their design, the Type 679 Manchester. In spite of the soundness of its design it was to be plagued by problems with its Vulture engines throughout its short service life. The Manchester, as well, was superseded by a development using four Merlin engines, the Type 683 Lancaster.

Although the Handley Page design remained basically unchanged, the design team, headed by G.R. Volkert, incorporated many alterations. The fuselage length and wingspan were increased, and the estimated all-up weight was increased from 26,300 lbs to 40,000 lbs. With the heavier engine weight forward of the wings, the centre of gravity was brought forward, and the sweep of the leading edge of the wing reduced from the planform of the HP.56. The rear spar was made straight, allowing for more room between the front and rear wing spars. Inboard of the inner engine nacelles, bomb bays were installed as were integral wing fuel tanks between the inner and outer engines. In order to accommodate these, the wing section was increased by twenty-one percent. These changes resulted in delays in the production schedule.

The smaller diameter of the propellers on the Merlin engines versus those proposed for the Vulture engines resulted in lower propeller efficiency, which gave them a lower cruise speed and range. Also, the drag was now increased because the larger wheels necessitated by the increase in gross weight could no longer be fully enclosed in the nacelles. Another factor that contributed to the drag was the thick wing section, which was twenty-one percent of the wing chord at the centre section tapering to nine percent at the wingtip. Although thick wing sections had been tested at the National Physics Laboratory and found to have low drag, these results were either faulty or had been improperly interpreted. By the time this had been determined, it was too late to change the design. A thinner wing section would have lessened the drag and increased performance, but would have made the housing of bombs in the wing bomb bays more difficult. The deep fuselage with its large frontal area also increased drag, but allowed fairly easy access over the wing spars and successful abandoning of a crashed aircraft. The night black finish which was applied to the undersides also contributed to drag, as did the non retractable landing lights.

In January 1938 construction began on the first prototype, serial number L7244, at the Cricklewood shops, and was completed just before the start of the war. It had been redesignated HP.57, and later allotted the name Halifax. The slab sided fuselage featured a two-gun nose turret and four-gun tail turret. The deep bomb bay extended the length of the fuselage centre section, and bomb cells were built into the inner sections of the wings between the main spars. The wing was mounted in the mid fuselage position, with dihedral on the outer panels, automatic slots on the outer leading

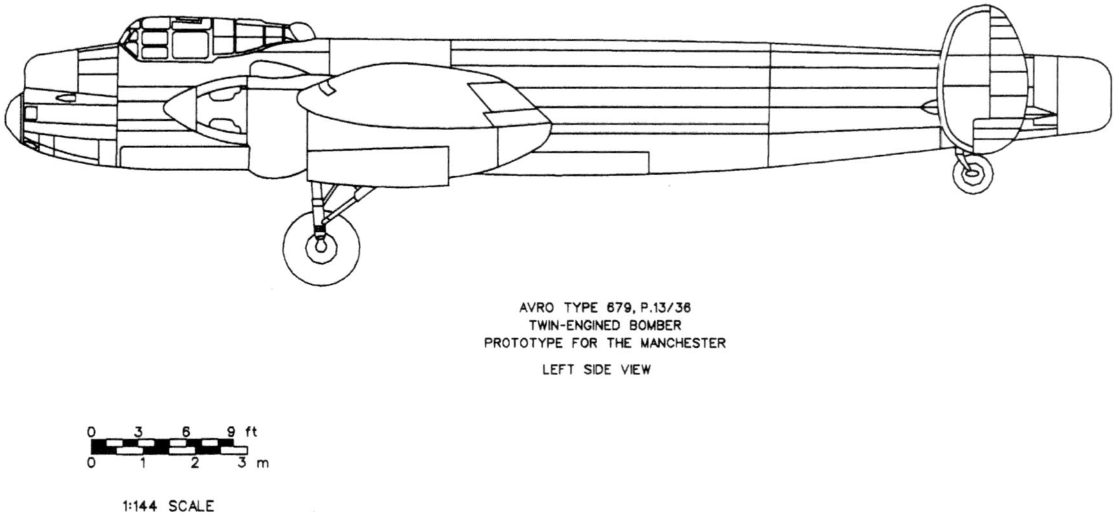

The Avro design proposal to Specification P.13/36, in a modified form, became the Avro Type 679 Manchester bomber. *(Andrew Tattersall)*

edges as on the HP.55 Hampden, and squared wing tips. The tailplane featured twin fins and rudders which acted as end plates. The original vertical stabilizers featured the swept back leading edge configuration, which later was proved to be the cause of loss of control and crashes.

The prototype aircraft emerged with no turrets installed, the turret positions being covered with metal fairings. DeHavilland three-blade metal propellers were installed, as were integral wing fuel tanks. Handley Page leading edge slots, designed to lower stall speed, were featured on the outer wing panels as they had been on its predecessor, the Handley Page Hampden. The direction finding loop antenna was unfaired, and no wireless telegraphy masts were installed for the high frequency radio.

The first flight was not conducted at the Handley Page facility at Radlett because of its limited size. The components were transported by road to the nearest non-operational RAF aerodrome, 13 OTU at Bicester, which had full sized hangars and all the facilities to reassemble and flight test the prototype.

Temporary fuel tanks were installed in the bomb bay for the first flight, since the integral wing fuel tanks were to be used to carry water ballast for full load and overload trials. It would have been necessary to dump the water ballast as quickly as possible in case of an emergency landing soon after takeoff, since these loads would have exceeded maximum permissible landing weight.

During taxiing trials, because the Lockheed hydraulic brakes were so slow acting, Major J.L.B.H. Cordes, the test pilot, refused to fly the aircraft until Dowty pneumatic brakes were installed. Although the changeover was accomplished by intense effort on the part of the Dowty team, there was not sufficient time to install an additional air compressor, so the system relied, initially, on a pair of compressed air bottles located in the fuselage. The first flight, with the undercarriage locked down, took place on 25 October 1939, piloted by Major Cordes. The aircraft was fully camouflaged with night black undersides. The aircraft performed well from the start, requiring little modification, although on a subsequent flight an elevator fractured and, fortunately, Major Cordes had enough control left to make a good landing. L7244 was flown at the Aircraft & Aeronautical Experimental Establishment (A&AEE) at Boscombe Down mainly for armament and equipment trials. It was grounded later in 1941 after it lost three of its propellers on takeoff, and became Instructional Airframe 3299M.

The second prototype, serial L7245, first flew in August 1940 from Radlett, again piloted by Major Cordes. It was complete with turrets and flew at an all up weight of 50,000 lbs. For comparison, the aircraft was equipped with constant speed Rotol propellers with Schwartz wooden blades, which, although not fully feathering, were chosen for production aircraft. It flew with fuel in the wing tanks and water ballast tanks in the bomb bay. The aircraft was camouflaged, but in accordance with the requirement for the finishing of experimental aircraft, the undersides were painted yellow with the serial number in large black characters underwing. After brief flight tests it was delivered to A&AEE at Boscombe Down. On 13 October 1940 L7245 narrowly escaped being written off when the starboard landing gear failed to retract fully after takeoff. Subsequent attempts to lock it in the up position caused the hydraulic accumulator to burst because the level of fluid was too high. Major Cordes lowered the gear by hand pump and executed a successful landing. The aircraft had dual controls installed, and continued as a trainer as Instructional Airframe 3474M until grounded in August 1942 when its engines became time expired.

The first production contract—No. 69649/37, Requisition 102/E11/37 for 100 aircraft—was placed in January 1939, and the first production aircraft, L9485, flew on 11 October 1940. The leading edge slats were installed on this aircraft, but were deleted from subsequent aircraft to allow for the fitting of barrage balloon cable cutters to the leading edges. L9485 served at Boscombe Down as an armament test aircraft until grounded as Instructional Airframe 3362M.

Production orders were issued to several subcontractors as well as the parent company. These included the English Electric Company, Fairey Aviation Co., Rootes Securities and the London Aircraft Production Group. This latter consortium was made up of motor coach and bus body builders including the LTPB, Park Royal Coach Works, Express Motor and Body Works, Chrysler Motors and Duple Bodies. Other than the English Electric Company, these firms had no previous experience in aircraft production. However, the aircraft structure consisted of fifteen main split sub-assemblies, which simplified its manufacture and assembly. Of these aircraft, about one in every hundred was flown to Handley Page at Radlett to have performance checks made by company pilots.

CHAPTER ONE

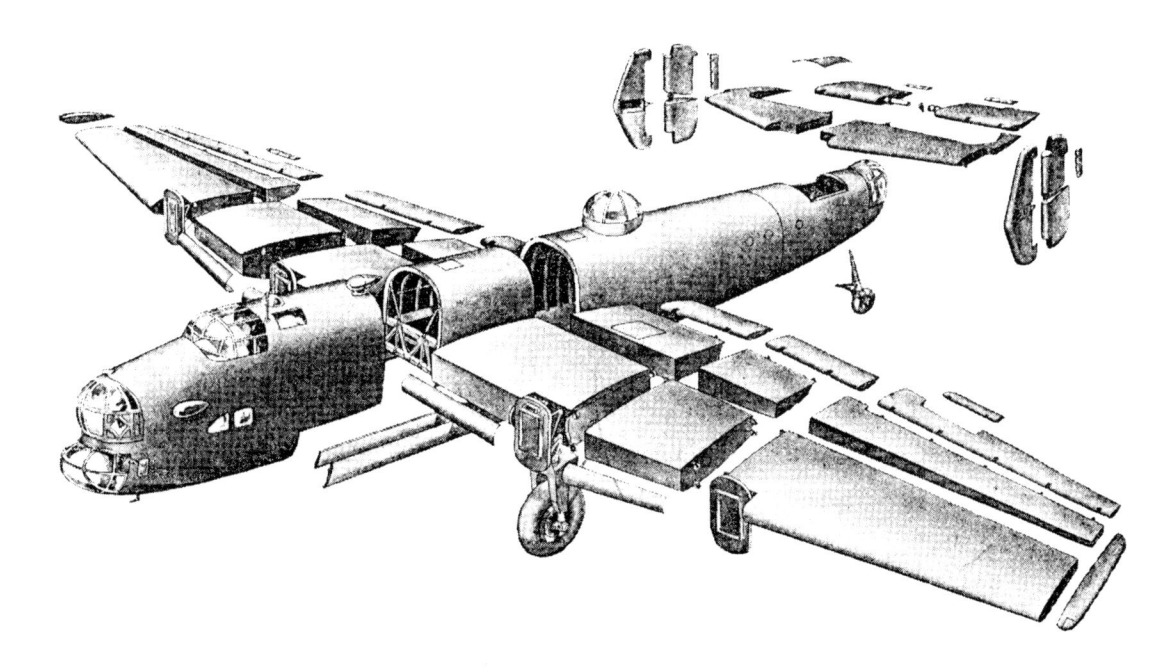

The Halifax B. Mk.I Series I, broken down into major component sub-assemblies. *(CF)*

On 11 October 1940 after brief flight tests at Radlett, the first production B.Mk.I, serial number L9485, went to Boscombe Down, and L9846 followed in December for crew familiarization. The first Halifax squadron, No. 35, had been formed on 5 November 1940 at Boscombe Down under the command of Wing Commander R.W.P. Collings. The squadron began training on the prototype L7244 which had been fitted with temporary dual controls. The unit moved to Leeming, Yorkshire and then to its operational base at Linton-on-Ouse, Yorkshire on 5 December. By March 1941, six crews had qualified. The first operational sortie was flown 10 March, and L9489 was misidentified and shot down on its return by a night defence fighter. In April, C Flight was expanded into No. 76 Squadron, officially formed on 1 May at Linton-on-Ouse before moving to its base at Middleton St. George.

The first fifty HP.57 B.Mk.I Series I were followed by twenty-five HP.57 B.Mk.II Series II with structural modifications permitting an increase in all up weight from 55,000 lbs to 60,000 lbs. The maximum permitted landing weight remained at 50,000 lbs, as did the fuel jettisoning requirements. These aircraft were also equipped with twin Vickers 0.303 calibre K guns in the side beam positions amidships to cover blind spots hidden from nose and tail turrets.

With this increased armament it was hoped that the Halifax could defend itself in day operations. However, on the first daylight raid by 35 Squadron on Kiel, Germany on 30 June, two of six aircraft were lost. On 24 July, five of fifteen aircraft from 35 and 76 Squadrons were lost in a raid on La Pallice, France. The Halifax was restricted to night operations for six months after these sorties.

Aircraft serial number L9600 carried the designation of HP.57 Halifax B.Mk.I Series III. From the seventy-sixth aircraft on, all were equipped with larger oil coolers to accommodate Rolls Royce Merlin XX engines, except the first nine which were equipped with Merlin X engines. Fuel capacity was again increased to 1,636 gallons, and an additional electric generator was added to the port outer engine. From serial number L9495 on, the T.1082/R.1083 radios were replaced by T.1154/R.1155 models. On 12 September 1941 L9608, the last B.Mk.I Series III, was officially named "Halifax" in a ceremony at Radlett by Viscountess Halifax, and later served in 76 Squadron RAF, coded MP*H.

The designation Halifax II was originally reserved for the HP.58, a heavily armed day bomber. It was to have been equipped with Boulton Paul Type H dorsal and Boulton Paul Type O ventral turrets, with twin 20mm cannon armament mounted amidships in a relatively low drag installation, similar to the arrangement in the Handley Page HP.52 Hampden. The rear fuselage was to be tapered to a pointed tail. A contract for one rear fuselage to be installed on prototype L7244 was placed on 25 July 1939, with the intention of phasing it into the production line after the tenth production aircraft. The mockup was inspected at the factory on 15 December 1939, but difficulties experienced in developing the turrets caused successive delays.

On 19 August 1940 Handley Page suggested that, to avoid further delays, the cannon armament be replaced by twin 0.50 calibre guns in dorsal and tail turrets, and that as a temporary measure, a dorsal turret with twin 0.303 guns and a tail turret with four 0.303 guns be installed. This matter was resolved by adding a twin 0.303 turret in the dorsal position, causing minimal problems on the production line. From the eighty-fifth aircraft on they were equipped with the Merlin XX engines and Boulton Paul Type C1 twin gun mid upper turret with the beam guns deleted. Fuel capacity was again increased to 1,882 gallons. These aircraft were designated HP.59 Halifax B.Mk.II Series I. The prototype, L9515, was first flown by Major Cordes at Radlett on 3 July 1941.

There were two further stages in armament development proposed by the Air Ministry. The first was that the four 0.303 gun Boulton Paul Type E tail turrets be replaced by a twin 0.50 calibre gun Boulton Paul Type D tail turret The second was that the twin 0.303 gun Boulton Paul Type C1 dorsal turret be replaced by the twin 20mm gun Boulton Type H turret. The cannon armed dorsal turret was never implemented. The installation of the twin 0.303 gun Boulton Paul Type C1 resulted in the P.59 Halifax B.Mk.II Series I. The installation of the twin 0.50 gun tail turret came about later, in 1944.

Although the Halifax had experienced heavy losses in early daylight raids due to the ineffectiveness of its small calibre defensive armament, it was proved in its role as a night bomber. The replacement of all Short

The cockpit of a HP.59 Halifax B.Mk.II Series I *(P. Vandersteen)*

CHAPTER ONE

Stirling and Avro Manchester aircraft in squadron service with the Halifax was proposed. However, Avro converted the twin Rolls Royce Vulture powered Manchester into the Avro Lancaster, powered by four Rolls Royce Merlins housed in the slim nacelles developed for the Bristol Beaufighter Mk.II. This compromise resulted in an aircraft with superior handling qualities much preferred by the aircrews.

The two prototypes and the first fifty production aircraft had been equipped with leading edge slats. These were found to contribute little to shorter takeoffs and landings, and were wired shut, and finally deleted on B.Mk.I Series III and subsequent aircraft. Armoured leading edges and balloon cable cutters were installed in their place .

In spite of the increased power of its engines, the performance of the Halifax B.Mk.II at full load was little better than the Halifax B.Mk.I. With its higher wing loading the lateral stability had deteriorated. The exhaust glow, easily visible from dead astern over the wing made it an easy target for cannon armed night fighters which could fire while still out of range of the 0.303 calibre tail turret guns. Improved 0.50 calibre and 20mm equipped turrets were virtually nonexistent.

Difficulties with landing gear were also experienced. The Messier tail wheels were designed to be retractable, but their self centering mechanism proved unreliable, and they were locked down to prevent damage in night landings when their position could not be checked visually. The Messier main landing gear was slow to retract, thus reducing takeoff performance, although it was otherwise reliable. It was also time-consuming to manufacture, causing delays in the accelerated production schedules. Dowty proposed a completely new levered suspension system, similar to that in use on the Manchester and Lancaster aircraft, which was quicker to operate and better suited to quantity production. On 11 April 1941 Halifax B.Mk.I, L9250, which was assigned to Rotol at Staverton for propeller de-icing trials, was used for trials with the Dowty undercarriage system. The Messier hydraulic system used castor oil based fluid (D.T.D 391) with natural rubber seals and glands, while the Dowty used mineral oil (D.T.D. 44) with synthetic rubber seals. Consequently it was not possible to achieve full compatibility and interchangeability of these systems without a long development and test program, even though the same geometry and structural pick up points were used.

At a meeting on 9 May 1941 with DTD, it was agreed that an improved hydraulic system would be developed for future Halifax production, using D.T.D.44 in Messier jacks with neoprene seals . This, however, would not relieve the immediate shortage of landing gear assemblies. The solution was to install Dowty undercarriage systems in production batches of Halifax aircraft, with a new Mark number, the HP.63 Halifax B.Mk.V. Otherwise identical to the B.Mk.II, L9250 flew as the B.Mk.V prototype in October 1941. Had the Dowty undercarriage used forgings as originally intended, it would have given good service. Instead a calculated risk was taken to use castings which took less time to manufacture, but they proved to be liable to brittle fracture at high stress levels. This was determined too late to change the production program. As a result, the maximum landing weight was restricted to 40,000 lbs. The B.Mk.V was to be restricted to meteorological reconnaissance and other coastal duties, and as tugs for training airborne troops and glider pilots. Six RCAF squadrons, however, operated them as bombers The first was No. 408(B) Squadron in October 1942, followed by No. 427(B), No. 428(B), No. 429(B), No. 431(B), No. 434(B). The B.Mk.V was also used as a bomber by RAF No. 76(B) Squadron, and two Free French squadrons, No. 346(B) and No. 347(B).

Production of the landing gear ceased in December 1943, and production of the B.Mk.V ended in January 1944 after 904 aircraft had been built. One aircraft, serial number DG399, was flown to Canada to be used as a pattern aircraft by Canadian Associated Aircraft to follow production of the Handley Page Hampden. This program was cancelled when the shortcomings of the Dowty undercarriage became evident.

Both HP.59 B.Mk.II and Hp.63 B.Mk.V aircraft underwent further improvements to performance. The first was the removal of the Boulton Paul Type C nose turret installation and its replacement by the "Tempsford" or "Z" nose fairing, and the removal of the Boulton Paul Type C1 mid upper turret. This would result in a weight saving of 1,450 lbs, and with the reduced drag, an increase in top speed of 16 mph. This was equivalent to a further saving of 840 lbs of fuel and oil over a range of 1,800 miles This modification was first discussed in a conference held by DTD in December 1941. Handley Page had long put forth the idea that these turrets were only of use in daylight operations and not worth the loss of performance at night.

This increase in performance was urgently needed by the Special Operations Executive for fast long range sorties across Europe, to supply the Polish Resistance with arms and equipment and to drop special agents. These modified aircraft were to be flown by No. 138 Squadron based at Tempsford. Three B.Mk.II aircraft were to be delivered as soon as possible. The first to be converted was W7774, flown on 15 August 1942. Conversion kits were produced at Rawcliffe repair depot and installed there on aircraft flown in from the squadrons. The conversion, Modification 398, resulted in aircraft being designated HP.59 B.Mk.II Series I (Special) or HP.63 B.Mk.V Series I (Special). Concurrently, the external fuel jettison pipes were deleted and improved flame damping exhausts were installed, without much initial success.

In the meantime, a new streamlined fuselage nose had been designed for future production. The first fairing, made in metal, was installed on B.Mk.I Series II, L9515 for aerodynamic testing. The slightly longer low drag nose provided a sitting position for the bomb aimer and increased stowage for radio and navigation equipment. The first transparent nose fairing, Modification 452, was installed at Hucknall on B.Mk.II Series I, W7814, which was written off on 23 September 1942. A second trial installation was made on DG276, and was approved for installation on both Halifax B.Mk.II and B.Mk.V. As well, a four-gun Boulton Paul Type A mid upper turret, was installed, which had far less drag than the Type C1 used on the B.Mk.II Series I. The turret installation on some aircraft featured a raised fairing like the one commonly found on the Lancaster. This allowed a 10 degree depression of the guns, but created a higher drag than an installation straight onto the upper fuselage decking, which allowed for no depression.

Aircraft equipped with this modification were designated HP.59 B.Mk.II Series IA, or HP.63 B.Mk.V Series IA. These also featured low drag nacelles derived from the Bristol Beaufighter Mk.IIF which had rectangular Morris block radiators in place of the Gallay twin drum radiators as installed in the HP.57 Halifax B.Mk.I. The fully modified prototype HP.59 Halifax B.Mk.II Series IA, HR679 first flew at Radlett on 24 December 1942. This aircraft was 10 mph faster than Halifax B,Mk.II W7922, even when the latter had its Boulton Paul Type C1 upper turret removed. The new nose installation featured a single handheld 0.303 calibre Vickers K gun, known as a "scare" gun.

On Coastal Command aircraft, this gun was replaced by a belt fed 0.50 calibre gun. These aircraft, built by Rootes, were converted by Cunliffe-Owen at Eastleigh with FN.64 ventral turrets—first installed on W7650 in September 1942—and 690-gallon long range fuel tanks installed in the bomb bay. Special radar and meteorological equipment was also installed, depending on the aircraft's operational role. The first GR.Mk.II Series IA was JP258 and the first B.Met Mk.II Series Ia was DG344. Aircraft JD212 had a trial installation of rocket projectile rails under the front fuselage but these were not adopted for service use. Twin 0.50 calibre remote control Browning guns were installed in ventral blisters in HR909 and NA137 for operational trials in place of FN.64 turrets. These guns were prone to overheating and were replaced by the single 0.50 calibre Preston-Green mounting, which was far more effective at long range.

Late-production Halifax B.Mk.II and B.Mk.V aircraft had Merlin 22 engines using 100 octane fuel at 7 psi boost, allowing an increased capacity to 65,000 lbs all up weight. These engines were also less prone to coolant leaks. A total of 299 HP.59 Halifax Mk.II Series IA variants were built at Cricklewood, nearly all equipped with Boulton Paul Type A Mk.VIII four-gun upper turrets. Removal of the mid upper turret had been unpopular with crews, and the opinion was that a Series IA aircraft with a Boulton Paul Type A mid upper turret installed at deck level had no more drag than a Series I (Special) fuselage without the mid upper turret.

There was also a series of modifications made to bomb bay doors which, originally, were made compatible to either H2S radomes or the ventral turret but could not enclose a bomb larger than 2,000 lbs. These metal skinned doors were replaced by faired wooden doors on steel carrier frames capable of enclosing 4,000-lb bombs. Trial Type B doors were made to suit aircraft V9985, but proved unacceptable. Original Type A doors were fitted with new hinges and side flaps, now redesignated Type D doors, and interchangeable with new Type C doors which were being made for the Halifax B.Mk.III but would not fit Halifax B.Mk.II or B.Mk.V aircraft. V9985 successfully dropped dummies of the 4,000-lb "blockbuster" bomb intended for B.Mk.III aircraft equipped with H2S radar. The first operational 8,000-lb "blockbuster," consisting of two 4,000-lb cases joined in tandem, was dropped by R9457 of No. 76 Squadron on the night of 10 April 1942 in a raid on Essen, Germany.

The H2S radome mockup was installed on W7711, and the first full scanner equipment installed in V9977

CHAPTER ONE

(which crashed early in 1942). It was replaced by W7808 and W7823. Further radar development was continued on W7851 and HR815/G, with the final service installation being approved on B.Mk.III, HX238.

The last major modification to be incorporated on Halifax B.Mk.II and B.Mk.V aircraft was the introduction of enlarged vertical tail surfaces to correct rudder stalling. This had caused spiral instability in the event that there were two engines dead on one side, causing many crashes on final approach. The original Type A rudders installed on the HP.57 Halifax B.Mk.I had proved to be overbalanced at large rudder deflections. This had been cured by a bulbous leading edge, Type B, which had caused hunting in straight and level flight. This was remedied by cutting the leading edge back by two inches and increasing the balance tab movement, Type C, but this caused trailing in a sideslip. Enlarged fins and rudders, symmetrical above the tailplane, Type E, had been designed for the Halifax B.Mk.III and B.Mk.IV, but these, again, were too overbalanced at large rudder deflections. Cutting back the rudder leading edge of the Type E as for the Type C, resulted in the Type F, which again trailed in sideslip. The Type F nose balance was cropped at the top and bottom edges, resulting in the Type G which was satisfactorily flown on DG281 in December 1942. Independent wind tunnel investigations led to a combination of Type C rudder with a rectangular fin of fifty percent greater area being evolved. When tested on DK145, it was found to have eliminated both overbalance and trailing. Designated as Type D, it was standardized on all production HP.61 Halifax B.Mk.III aircraft. This modification was also a retroactive installation on all late Merlin powered aircraft, particularly those equipped with Merlin 22 engines and four-blade propellers or with H2S radar installation.

The total number of Merlin-engine-powered HP.57 Halifax Mk.I, HP.59 Halifax Mk.II and HP.63 Halifax B.Mk.V aircraft was 1,967. Of these, 701 were manufactured by Handley Page at Cricklewood, including the two prototypes, all B.Mk.I, and the remainder, B.Mk.II. English Electric built 900 B.Mk.II at Preston, and 450 B.Mk.II were built by the London Aircraft Production Group at Leavesden. Twelve B.Mk.II and 658 B.Mk.V were built by Rootes at Speke, and 246 by Fairey Aviation at Errwood Park.

In an effort to use the Halifax as a day bomber, the Air Staff approached Handley Page to develop a high altitude version powered by either Rolls Royce Merlin 60 Series or Bristol Hercules VIII engines. Both engines were equipped with two-stage superchargers offering maximum power rating at 30,000 ft altitude. The Hercules VIII engine was plagued by incurable supercharger surging and was abandoned, but the Merlin 60 Series engines had gone into production and were proposed for the HP.60A Halifax B.Mk.IV.

This variant was to have had a strengthened bomb floor and enlarged bomb bay doors to allow the carrying of 4,000-lb and 8,000-lb bombs. The wingspan was increased to 104 ft by the addition of extended wing tips and the enlarged symmetrical Type E fin and rudder assemblies. Halifax HR756 was earmarked for the prototype. Although built to be equipped with Rolls Royce Merlin 60 series engines, only Merlin XX were available for initial flight testing in March 1943. In this form the variant was designated HP.59 B.Mk.II Series 2. The intention was to install Merlin 60 engines when available, to bring it up to Halifax B.Mk.IV standard. The aircraft was allotted to Rolls Royce at Hucknall for engine development programs. The extended wing tips were never installed, although the aircraft flew with both Merlin 61 and 65 Series engines.

On 5 August 1941 Handley Page had recommended to the Controller of Research and Development that certain features of the Halifax B.Mk.IV be introduced into production aircraft. In September the Controller responded with a request for the introduction of Bristol Hercules VI or XVI engines, with structural modifications to raise the all up weight to 64,000 lbs. This variant emerged as the HP.61 Halifax B.Mk.III. It was estimated to achieve 307 mph at 21,000 ft. Hercules 100 engines would be installed later in place of the failed Hercules VIII engines. Handley Page had already been requested to incorporate several modifications which would have seriously interfered with production schedules. Although they agreed with the installation of the Boulton Paul Type A mid upper turret and FN.64 ventral turret for daylight operations, the company suggested deferring the other modifications until the 200th aircraft was built. These included provisions for target towing and paratroop drops, heavy gauge protective engine cowlings, multiple flare chutes, revised ammunition tracks to the tail turret to clear the paratroop exit, a revised dual control pilot seat, rocket gear for short takeoff, and arrester gear for landing. These modifications would certainly affect the efficiency of the aircraft as a bomber. The Controller replied with a deferral of all

Handley Page Halifax

outstanding modifications except exhaust flame dampers, engine rear armour, astrograph and D/F loop, fixed fittings for 4,000- and 8,000-lb bombs, increased electricity generation and provision for glider towing.

In January 1942 Halifax B.Mk.II Series I (Special) R9534, was selected for the installation of the Hercules VI engines, flying for the first time on 12 October 1942 as the HP.61 Halifax B.Mk.III prototype. In December, the Boulton Paul Type C1 mid upper turret was replaced by the Type A. Mk.III turret, and further improvement in performance was achieved by retracting the tailwheel and fitting Beaufighter type Hercules engine cowlings. Early production aircraft were fitted with the single 0.50 calibre Preston-Green ventral mounting, while later production aircraft had H2S radar installed in its place, and had extended wing tips. The first true Halifax B.Mk.III, HX226, the 700th production aircraft, was flown in August 1943. More HP.61 B.Mk.III aircraft were built than any other variant, with production being shared by all five main assembly groups, amounting to 2,127 aircraft. The first production aircraft were issued to 433(B) Squadron RCAF, and 466(B) Squadron RAAF in November 1943. Forty-one operational squadrons and other units were eventually equipped with this variant.

By February 1944 the Hercules 100 engine, as a self contained "power egg," became available, offering increased performance using 100 octane fuel in conjunction with RAE-Hobson injection carburetors. These required a high pressure fuel supply which featured a completely revised fuel system with seven flexible fuel cells in each wing, arranged in inboard and outboard groupings. Each engine was supplied by adjacent fuel cells and the fuel cocks were relocated to the flight deck from behind the front wing spar. The fixed fuel capacity was 2,190 gallons with an additional 690 gallons available in three auxiliary fuel tanks located in the fuselage bomb bay. With this fuel system, larger Gallay oil coolers, the Gravely cabin heating system using four exhaust system heaters, and all retrospective modifications including the extended wing tips, the variant was designated HP.61 Halifax B.Mk.VI. When airframe production outstripped the availability of Hercules 100 engines, Hercules XVI engines were installed. These aircraft were designated HP.61 Halifax B.Mk.VII.

The Hercules 100 engines were first installed at Filton in the trial Halifax B.Mk.III, HX234, and in the prototype B.Mk.VI, LV776, which was first flown on 19 December 1943. The first production B.Mk.VI was NP715, first flown on 10 October 1944. This aircraft became a trial installation aircraft at Radlett after surviving months of operational service. In 1946/47 it was used to test a lateral control system using curved spoilers moving in and out of slots immediately ahead of the ailerons in the upper surface of the wings.

The HP.61 Halifax B.Mk.VI and B.Mk.VII were the last bomber variants produced. B.Mk.VI production totalled 457 aircraft, while 423 B.Mk.VIIs came off the assembly lines, and others were cancelled after VJ Day. Most of these aircraft were fully "tropicalized" and many operated in South East Asia Command against the Japanese.

The Halifax program was constrained by the two conflicting demands of quantity production and the necessity to maintain a maximum output of aircraft with the fewest possible major design changes. In spite of this, research was pursued into achieving higher performance and greater efficiency. However, in retrospect, it would seem apparent that the delays in refining the Halifax design with a new streamlined fuselage nose and low drag Type A.Mk.VIII turret would have been obvious ways of achieving improved performance from the outset. In spite of these shortcomings, the aircraft was structurally robust and survived extensive damage by flak and night fighters. Finally, although the Rolls Royce Merlin-equipped earlier versions were underpowered and suffered marginal performance, the later Hercules-powered aircraft were greatly improved and in many respects were the equal of the Avro Lancaster, although the Lancaster proved to have a better load carrying ability.

CHAPTER TWO

The Halifax in Canadian Service

The nose section of a Handley Page HP.61 Halifax B.Mk.III running up prior to takeoff. Note the .303 calibre "scare gun" in the nose perspex. *(CF PL-43476)*

Handley Page Halifax

The Handley Page Halifax bomber became the mainstay of the Canadian Squadrons in the Royal Air Force Bomber Command. Although some squadrons were initially equipped with variants of the Vickers Wellington or the Handley Page Hampden, all fifteen of them were equipped with the Halifax at some point in their wartime service, and some were formed on and operated only the Halifax. As well, three Royal Canadian Air Force Heavy Conversion Units were equipped with the Halifax to convert graduates of the operational training units to the Halifax heavy bomber.

The Halifax, although designed as a heavy bomber, also served in a variety of roles such as maritime reconnaissance and antisubmarine operations, glider tug, paratroop dropping, meteorological reconnaissance, air ambulance and freighter, and on special operations, dropping agents and supplies to resistance groups in enemy occupied areas.

The first Canadian squadron to operate the Halifax was No. 405 (B) "Vancouver" Squadron RCAF, which had formed in April 1941 at Driffield, Yorkshire. Operating in No. 4 Group of RAF Bomber Command, 405 (B) was equipped with the Vickers Wellington B.Mk.II medium bomber. The unit moved to Pocklington, Yorkshire in June 1941 and the changeover to the Halifax began in January 1942 when a conversion flight was formed, equipped with Halifax B.Mk.I Series I and B.Mk.II Series I aircraft. The conversion unit was disbanded on 7 October 1942 and absorbed into 1659 Heavy Conversion Unit.

The squadron stood down on 17 April 1942 pending re-equipment with Handley Page HP.59 Halifax B.Mk.II Series I aircraft. From February to August 1942 405(B) Squadron was commanded by W/C J.E. Fauquier, the first Canadian to command a heavy bomber squadron. By 25 May sixteen crews were qualified on the new bombers. By 29 May sixteen of the seventeen aircraft on strength were operationally ready. On the night of 30/31 May, sixteen crews were briefed on the bombing of Cologne, Germany, in Bomber Command's first 1,000-Bomber Raid. Of the fifteen crews that were airborne, ten succeeded in bombing the target, and three experienced mechanical problems and returned early. One aircraft, after taking off late, bombed Dusseldorf, Germany. Aircraft LQ*K, W7707, flown by Sgt. Wadman failed to return from the operation, and was believed to have suffered a midair collision with a Lancaster of No. 61 Squadron, RAF.

Once the maintenance crews gained experience on the Halifax, the aircraft proved to have a more reliable electrical system than the Wellingtons previously flown by the squadron. Problems that had been experienced with the R/T and intercom on the Wellingtons were almost nonexistent on the Halifax. However, problems were experienced with the engine coolant radiators developing leaks.

As well as the aircraft lost on the night of 30/31 May, 405(B) Squadron lost nine more crews on twelve operations during June. On the night of 8/9 June, three of eleven aircraft were lost on operations to Essen, Germany. Aircraft LQ*H, W7708, flown by F/L A. MacLean was coned by searchlights while approaching the target and a near miss by antiaircraft guns threw the aircraft onto its back. The pilot managed to right the aircraft, but found the ailerons jammed, causing the aircraft to fly in a circle. The bombs were jettisoned, and after the crew managed to free the controls and the pilot set course for home, the aircraft was attacked by a Messerschmidt Bf.110 night fighter. The crew returned fire, sending the fighter spinning down in flames, but the Halifax had been seriously damaged, with two port engines out of commission. The crew was ordered to bail out from low altitude and the Halifax crashed near Heteren, Germany. F/L MacLean, after surviving a heavy landing by parachute, was able to evade capture and after three months reached Gibraltar, and returned to England. He was awarded the DFC, but was taken off operations because of his valuable knowledge of the escape organization in occupied Europe. He survived the war and later became the Premier of Prince Edward Island.

In August 405(B) Squadron moved to Topcliffe, Yorkshire. Losses continued on operations, not only to enemy action, but to mechanical failure as well. Since beginning operations on the Halifax at the end of May, thirty-three crews had been lost.

In October the squadron was assigned to operations with 19 Group RAF Coastal Command, and moved to Beaulieu, Hampshire. From this base, on four months of operations, the squadron flew patrols over the Atlantic on escort to the convoys transporting men and equipment for Operation Torch, the Allied invasion of North Africa. There were several encounters with U-boats during this time, although no successful attacks were made. Operations against the French ports were also flown during this period, and aircraft were damaged in attacks by Arado 196 float plane fighter aircraft. On 9 December aircraft LQ*N, flown by Sgt. Burton was attacked by two Junkers 88 fighters while shadowing an

CHAPTER TWO

enemy convoy, and his mid upper gunner managed to shoot both down. On another patrol on 4 January, F/O W. W. Colledge was attacked by a flight of four Junkers 88 fighters and his mid upper gunner managed to destroy one and damage a second.

From 4 to 12 January, all 405(B) Squadron aircraft were grounded after several aircraft experienced engine problems. These problems were remedied after investigation by engineering staff. Some aircraft also experienced loss of all four engines at altitude, some descending to as low as 1,000 ft before regaining control. Many may have been lost after experiencing this problem. The possible cause was in the high tension ignition system. As well, the Halifax aircraft had a low service ceiling and was vulnerable to flak as a consequence.

The last operational sortie by the squadron as a part of Coastal Command was made by four aircraft on 27 February 1943. Beginning on 1 March, the unit returned to Topcliffe, Yorkshire to take up operational duties with No. 6 Group (RCAF). The first operation was flown on 11 March, in a raid on Stuttgart, Germany. On that same day, orders were received transferring the squadron to Leeming, Yorkshire, and the last operation from Topcliffe was flown the next day, from which all aircraft returned. The stay at Leeming was short-lived.

On 8 April 1943 the squadron was ordered moved to No. 8 Pathfinder Force Group, RAF Station Gransden Lodge. The effective date of transfer was 19 April, the day on which command of the squadron was taken over again by W/C J.E. Fauquier, DFC. As part of Pathfinder Force, 405(B) Squadron took on the task of locating and marking the target for the following force, as well as bombing the target. This group was staffed by the elite of the experienced bomber crews who were equipped with the latest technological innovations. The first operation was 26 April against Duisburg, Germany. The squadron was the only unit in the Pathfinder Force flying Halifax aircraft, at the preference of the squadron commander. However, the changeover to the Lancaster began in August 1943, and the last Halifax aircraft was replaced by the end of September. The Halifax B.Mk.II aircraft had served with 405 Squadron for some fifteen months.

No. 408(B) "Goose" Squadron was the second RCAF bomber squadron to operate the Halifax heavy bomber. Formed on 24 June 1941 at Lindholme, Yorkshire, and operating in No. 5 Group of RAF Bomber Command, the unit was originally equipped with Handley Page Hampden B.Mk.I aircraft. The squadron moved to Syerston, Nottinghamshire in July, and to Balderton, Nottinghamshire in December of that year. Between January and March of 1942, the squadron was relocated to North Luffenham, Nottinghamshire while the runways at Balderton were being resurfaced.

On 16 May 1942 408(B) Conversion Flight was formed at Syerston to introduce the Avro Manchester B.Mk.IA into service, but was cancelled on 19 June 1942 when the performance of the aircraft proved unacceptable. The flight was reformed on 20 September 1942 at Leeming, Yorkshire with Handley Page Halifax B.Mk.I, B.Mk.II, and B.Mk.V aircraft. The conversion flight was disbanded on 7 October 1942 and was absorbed into 1659 Heavy Conversion Unit.

The squadron moved once again, in September 1942, this time to Leeming to operate under No. 4 Group of RAF Bomber Command. The squadron stood down on 7 November 1942 to undergo conversion to the Handley Page Halifax. Although the unit was initially equipped with Halifax B.Mk.V aircraft, Halifax B.Mk.II Series I were taken on strength in December 1942. On 9 December 1942 six squadron pilots went on familiarization trips in Halifax aircraft of 102(B) Squadron RAF on a raid to Turin, Italy, and again on 20 December to Duisburg, Germany, all returning safely. During conversion one crew had been lost and three aircraft written off. By 9 January 1943 the squadron was again operational.

In March 1943 Bomber Command focused on deep penetration raids into Germany utilizing the greater range of the heavy bombers. However, losses increased as the German defence system deployed night fighters against both the incoming and returning bomber streams. In August 408(B) Squadron was notified of a move to Linton-on-Ouse for conversion to Avro Lancaster B.Mk.II aircraft, thus ending its first period of service with the Halifax aircraft.

In September 1944 408 was again equipped with the Handley Page Halifax, this time with the Hercules-powered B.Mk.III and B.Mk.VII aircraft with significantly improved performance. The last operational mission with Halifax aircraft took place on 25 April 1945 when gun emplacements on Wangerooge Island were bombed.

No. 419(B) "Moose" Squadron was formed in England on 15 December 1941 at Mildenhall, Suffolk, operating in No. 3 Group of RAF Bomber Command, and the unit was equipped with Vickers Wellington B.Mk.IC and B.Mk.III aircraft. The squadron relocated to Leeming, Yorkshire, as part of No. 4 Group in August 1942. Further moves were made in Yorkshire to Topcliffe, later in August, and to Croft in October, 1942. That month the squadron was re-equipped with Handley Page HP.59 Halifax B.Mk.II Series I heavy bomber aircraft, and a further move was made to Middleton St. George in Durham. The squadron was to operate the Halifax B.Mk.II until it was replaced by the Canadian built Avro Lancaster B.Mk.X in April 1944.

No. 420(B) "Snowy Owl" Squadron was formed on 19 December 1941 at Waddington, Lincolnshire, and operated in No. 5 Group of RAF Bomber Command. The unit was equipped with Handley Page Hampden B.Mk.I aircraft. A conversion flight was formed in April, 1942 to introduce the Avro Manchester B.Mk.IA into service, but, as with 408(B) Squadron, the flight was cancelled in June when the performance of the aircraft proved unacceptable. The squadron moved to Skipton-on-Swale and No. 4 Group in August 1942 and was re-equipped with Vickers Wellington B.Mk.III, and later B.Mk.X aircraft. A further move was made in October 1942 to Middleton St. George, Durham where in January 1943 420(B) joined the newly formed No. 6 (RCAF) Group.

On 25 June 1942 425(B) "Alouette" Squadron was formed at Dishforth, Yorkshire and operated in No. 4 Group. The unit was equipped with Vickers Wellington B.Mk.III aircraft. It was designated "French-Canadian," and both air and ground crew of French descent were selected from throughout Bomber Command. The squadron flew its first operational mission on 5 October 1942 when five aircraft bombed Aachen, Germany.

No. 424(B) "Tiger" Squadron was formed on 15 October 1942 at Topcliffe, Yorkshire, and operated in No. 4 Group RAF Bomber Command. The unit was equipped with Vickers Wellington B.Mk.III medium bombers.

The same day, No. 426(B) "Thunderbird" Squadron RCAF was formed at Dishforth, also operating in No. 4 Group, also equipped with Vickers Wellington B.Mk.III medium bombers. On 14 January 1943 No. 426(B) "Thunderbird" Squadron flew its first operational mission when seven Wellington B.Mk.III aircraft bombed Lorient, France.

On 7 November 1942 three more squadrons were formed, all equipped with Vickers Wellington B.Mk.III medium bombers. At Croft, Yorkshire, 427(B) "Lion" Squadron RCAF operated in No. 3 Group. The first operational mission was flown on 14 December 1942 when three aircraft were despatched on a mine laying mission in the area of the Frisian Islands. On 5 May 1943 the squadron moved to Leeming and was converted to Handley Page HP.63 Halifax B.Mk.V (Special) heavy bombers. In January 1944, the squadron re-equipped with Handley Page HP.61 Halifax B.Mk.IIII aircraft. The last operational mission on Halifax aircraft took place on 3 March 1945, in the form of mining off the coast of Norway.

No. 428(B) "Ghost" Squadron was formed at Dalton, Yorkshire, operating in No. 3 Group. The first operational mission was flown on 26 January 1943 when six aircraft set out to bomb U-boat pens at Lorient, France. In April 1943 the unit converted to Vickers Wellington B.Mk.X aircraft. In June the squadron relocated to Middleton St. George, Durham, where it converted to the Handley Page HP.63 Halifax B.Mk.V heavy bomber. In September, Handley Page HP.59 Halifax B.Mk.II aircraft were added to the operational inventory. In January 1944 conversion began to Handley Page HP.61 Halifax B.Mk.III aircraft, but this was suspended when it was decided to convert to the Avro Lancaster. The last operational mission on Halifax aircraft was flown in June 1944.

No. 429(B) "Buffalo" Squadron was formed at East Moor, Yorkshire, operating in No. 3 Group. In January 1943 the squadron added Vickers Wellington B.Mk.X to their operational inventory. The squadron flew its first operational mission on 21 January when three Wellington B.Mk.X aircraft laid mines off Terschelling, and flew its first bombing mission on 26 January when ten aircraft bombed Lorient, France. In time the unit converted to the Handley Page HP.59 Halifax B.Mk.II heavy bomber. A gradual conversion to the HP.63 Halifax B.Mk.V, which began in November 1943 was suspended, and the Handley Page HP.61 Halifax B.Mk.III was taken on strength beginning in January 1944. The last Halifax operational mission took place on 15 March 1945 against the Castrop-Rauxel oil refinery.

No. 431 "Iroquois" Squadron formed at Burn, Yorkshire on 11 November 1942 and operated in No. 4 Group. The unit was equipped with Vickers Wellington B.Mk.X medium bombers. The first operational mission was on 2 March 1943 when seven aircraft were despatched to lay mines in the area of the Frisian Islands, and its first operational bombing mission took place on 5 March when three aircraft took part in a raid on Essen, Germany.

On 15 July 1943 the unit moved to Tholthorpe, Yorkshire, converting at the same time to the Handley Page HP.63 Halifax B.Mk.V heavy bomber. On 10 December 1943

CHAPTER TWO

the squadron moved to Croft in Yorkshire, remaining on that base to the end of the war. Conversion to Handley Page HP.61 Halifax B.Mk.III began in March 1944, although the B.Mk.V aircraft continued in service until July. The last Halifax operational mission was flown on 25 October 1944.

On 25 October 1942 the No. 6 (RCAF) Group, RAF Bomber Command was formed at Linton-on-Ouse, Yorkshire, and on 6 December the unit moved to permanent headquarters at Allerton Park, Yorkshire. Effective 0001 hours on 1 January 1943 the eleven RCAF bomber squadrons were incorporated into the new No. 6 (RCAF) Group and declared operational. No. 405(B) Squadron, although still on assignment to No. 19 Group RAF Coastal Command was among the eleven, as were the other two Halifax equipped squadrons, 408(B), 419(B) and the new RCAF bomber squadrons that had formed during 1942, equipped with Vickers Wellington medium bombers.

Mine laying, or "gardening," was a major activity in the early part of 1943. With the Battle of the Atlantic nearing its peak, main targets were the submarine pens at Lorient and St. Nazaire, France.

On 1 March 1943 No. 61 Training Base was formed at Topcliffe, controlling RCAF Stations Topcliffe, Dishforth and Dalton in Yorkshire. On 14 March, No. 1659 Heavy Conversion Unit was moved to Topcliffe from Leeming where it had formed on 6 October 1942. No. 1664 "Caribou" Heavy Conversion Unit formed at Croft on 10 May 1943, and moved to Dishforth on 7 December 1943. No. 1666 "Mohawk" Heavy Conversion Unit formed at Dalton on 15 May 1943, and moved to Wombleton, Yorkshire on 21 October 1943.

No. 432(B) "Leaside" Squadron was formed at Skipton-on-Swale on 1 May 1943, operating in No. 6 (RCAF) Group. The unit was equipped with Vickers Wellington B.Mk.X medium bombers. The first operational mission was flown on 23 May 1943 when ten aircraft were despatched to bomb Dortmund, Germany. The unit converted to Avro Lancaster B.Mk.II aircraft in October 1943. In February 1944, the squadron converted to Handley Page HP.61 Halifax B.Mk.III aircraft, and to HP.61Halifax B.Mk.VII in June 1944.These aircraft were operated until the squadron disbanded on 15 May 1945. The last operational mission was flown on 25 April 1945 against gun emplacements on Wangerooge Island.

No. 434(B) Squadron was formed on 13 June, 1943 at Tholthorpe, Yorkshire and operated in No. 6 (RCAF) Group. The unit was equipped with Handley Page HP.63 Halifax B.Mk.V heavy bombers. The first bombing mission was flown on 12 August when ten aircraft carried out a raid on Milan, Italy. The squadron moved to Croft on 11 December 1943. The unit began conversion to Handley Page HP.61 Halifax B.Mk.III in May 1944. The last operational mission with Halifax aircraft was flown on 21 December 1944 against Cologne, Germany.

In May 1943 three squadrons were posted to North Africa as part of 331 (RCAF) Wing, No. 205 Group in Mediterranean Air Command. From bases in Tunisia, 420(B), 424(B), and 425(B) Squadrons flew tropicalized Vickers Wellington B.Mk.III and B.Mk.X aircraft from June to October in support of the invasions of Sicily and Italy. All three returned to their bases in England in November 1943.

On 1 June 1943 Linton-on-Ouse Operational Base was formed at Linton-on-Ouse, controlling RCAF Stations Linton-on-Ouse, East Moor, and Tholthorpe, in Yorkshire. It was redesignated No. 62 "Beaver" (Operational) Base on 6 October 1943. Two further No. 6 (RCAF) Group bases were formed on 1 May 1944; No. 63 (Operational) Base at Leeming, controlled RCAF Stations Leeming and Skipton-on-Swale, and No. 64 (Operational) Base at Middleton St. George, Durham, controlled RCAF Stations Middleton St. George and Croft.

On 25 September 1943 No. 433(B) "Porcupine" Squadron formed at Skipton-on-Swale, operating in No. 6 (RCAF) Group. It was equipped with Handley Page HP.61 Halifax B.Mk.III aircraft, which first arrived at the squadron on 3 November. The first B.Mk.III production aircraft were issued to 433(B) Squadron RCAF in November 1943. The first operational mission was flown on 2 January 1944, when four aircraft were despatched to lay mines in the area of the Frisian Islands. The last operation flown with Halifax aircraft was on 16 January 1945.

In November 1943 the three squadrons that had flown in the Mediterranean theatre returned to No. 6 (RCAF) Group, operating Vickers Wellington B.Mk.X aircraft. No. 420(B) Squadron was based at Dalton. When, in December, the unit relocated to Tholthorpe a conversion was made to Handley Page Halifax B.Mk.III, and later B.Mk.VII aircraft.

No. 424(B) Squadron, was based at Skipton-on-Swale. In December, conversion was made to Handley Page Halifax B.Mk.III aircraft, which were flown until January 1945. The unit then converted to Avro Lancaster B.Mk.I/III aircraft.

Handley Page Halifax

No. 425(B) Squadron was based at Dishforth. In December, when the unit relocated to Tholthorpe, a conversion was made to Handley Page Halifax B.Mk.III, and later B.Mk.VII aircraft, which were flown until the hostilities in Europe ended in May 1945.

The fifteenth Canadian bomber squadron was formed using trained personnel from within No. 6 (RCAF) Group. Serving from its inception in Coastal Command as No. 415(TB) "Swordfish" Squadron, the unit reformed on 12 July 1944 as No. 415(B) "Swordfish" Squadron. Relocated to East Moor, the squadron was equipped with Handley Page Halifax B.Mk.III and later, some Handley Page B.Mk.VII aircraft. The first operational mission was carried out on 28 July 1944 when aircraft were despatched to bomb Hamburg, Germany. The last operational mission took place on 25 April 1945 when eighteen aircraft bombed gun positions on Wangerooge Island.

By the end of the war, only 415(B) "Swordfish" Squadron and 432(B) "Leaside" Squadron were still equipped with Handley Page Halifax bombers, and were disbanded at their bases. The remainder had either converted to Canadian built Avro Lancaster B.Mk.X aircraft and had returned to Canada in No. 6 (RCAF) Group, Eastern Air Command for training as part of RAF Tiger Force, or joined No. 1 Group RAF Strike Force equipped with Avro Lancaster B.Mk.I/II aircraft. The Halifax, although in some respects inferior to the Avro Lancaster, had served with distinction in the Canadian Squadrons of Bomber Command.

CHAPTER THREE
Aircraft Description and Drawngs

Handley Page HP.57 Halifax B.Mk.I Series I heavy bomber (ZA*E of No. 10 Squadron RAF). This variant served only with the Heavy Conversion Flights of No. 405(B) Squadron and 408(B) Squadron, which amalgamated to form 1659 Heavy Conversion Unit. *(CF PL-8916)*

Handley Page Halifax

VARIANTS POWERED BY ROLLS ROYCE MERLIN V-12 ENGINES

Handley Page HP.57 Halifax B.Mk.I Series I

Description: All-metal, mid wing, cantilever monoplane heavy bomber, equipped with Messier undercarriage and fixed tailwheel.

Powerplant: Four Rolls Royce Merlin X V-12 cylinder engines rated at 1,130 hp with Delaney Gallay coolant radiators, driving three-blade Rotol constant speed compressed wood propellers.

Dimensions:
- Height: 20 ft. 9 in.(6.32 m) (tail down, including DF loop antenna)
- Length: 70 ft. 1 in. (21.36 m)
- Wingspan: 98 ft. 8 in. (30.07 m)
- Wing Area: 1,250 sq. ft. (116.13 m2)
- Tailplane Span: 30 ft. 4 in. (9.25 m)
- Undercarriage Track: 24 ft. 8 in. (7.52 m)
- Dihedral Inner Mainplane: 00
- Dihedral Outer Mainplane: 20

Weights: Empty weight: 33,720 lbs. (15,296.4 kg)

Max takeoff weight: 55,000 - 60,000 lbs. (24,948 - 27,216 kg.)

Performance: Max speed: 255 mph at 7,000 ft. (410.4 kph at 2133.6 m)

Economical Cruising Speed: 195 mph at 15,000 ft. (313.9 kph at 4572 m)

Takeoff Distance: 3060 ft. at 55,000 lbs. to clear 50 ft. obstacle.
(932.7 m at 22,680 kg to clear 15.24 m obstacle)

Landing Distance: 2,550 ft. (777.3 m)

Initial Rate of Climb: 750 ft/min. at 58,000 lbs. all up weight
(228.6 m/min at 26,308.8 kg. all up weight)

Service Ceiling: 18,000 ft. @ maximum weight
(6,264 m @ maximum weight)

Normal Range: 1,860 miles with full tanks and 5,800 lbs. bomb load
(2,993 km with full tanks and 2,631 kg. Bomb load)

Armament: Boulton Paul Type C Nose Turret with two Browning Mk.II 0.303 machine guns and 1,000 rounds of ammunition per gun.

Twin Vickers K 0.303 machine guns on each side of the rear fuselage in beam position.

Boulton Paul Type E Tail Turret with four Browning Mk.II 0.303 machine guns and 2,500 rounds of ammunition per gun.

Bomb Load: *Maximum 13,000 lbs.*

One 500 lb. bomb in each of six wing bomb bays
Two 2,000 lb. bombs + six 1,000 lb. bombs in fuselage bomb bay

Or:

One 500 lb. bomb in each of six wing bomb bays
Four 2,000 bombs in fuselage bomb bay

Or:

One 500 lb. bomb in each of six wing bomb bays
Two 1,500 lb. mines + six 500 lb. bombs in fuselage bomb bay

Or:

One 500 lb. bomb in each of six wing bomb bays
Nine 500 lb. bombs in fuselage bomb bay

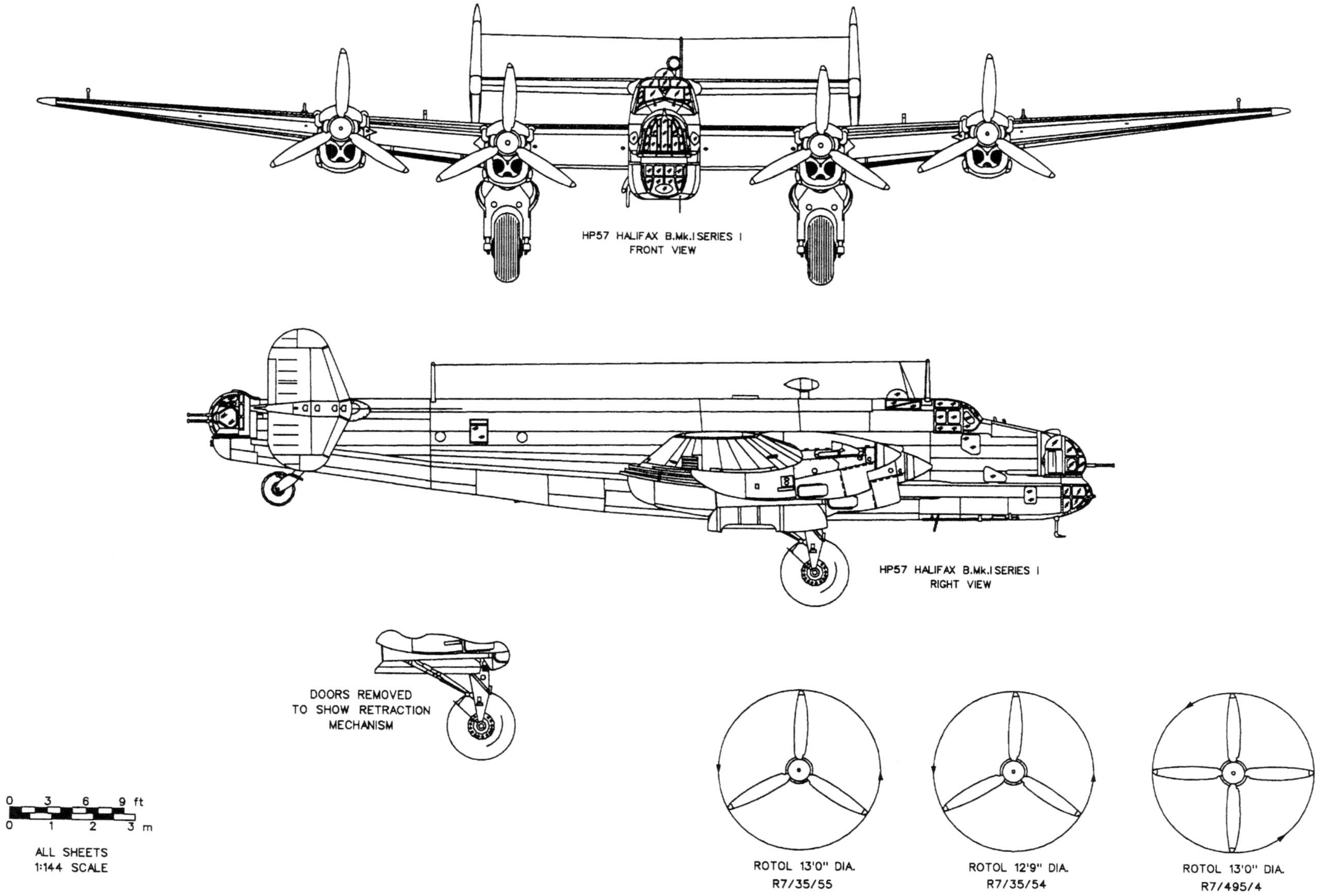

Front and right side view of Handley Page Halifax B.Mk.I Series I. *(Andrew Tattersall)*

Left side view and sections of Handley Page Halifax B.Mk.I Series I *(Andrew Tattersall)*

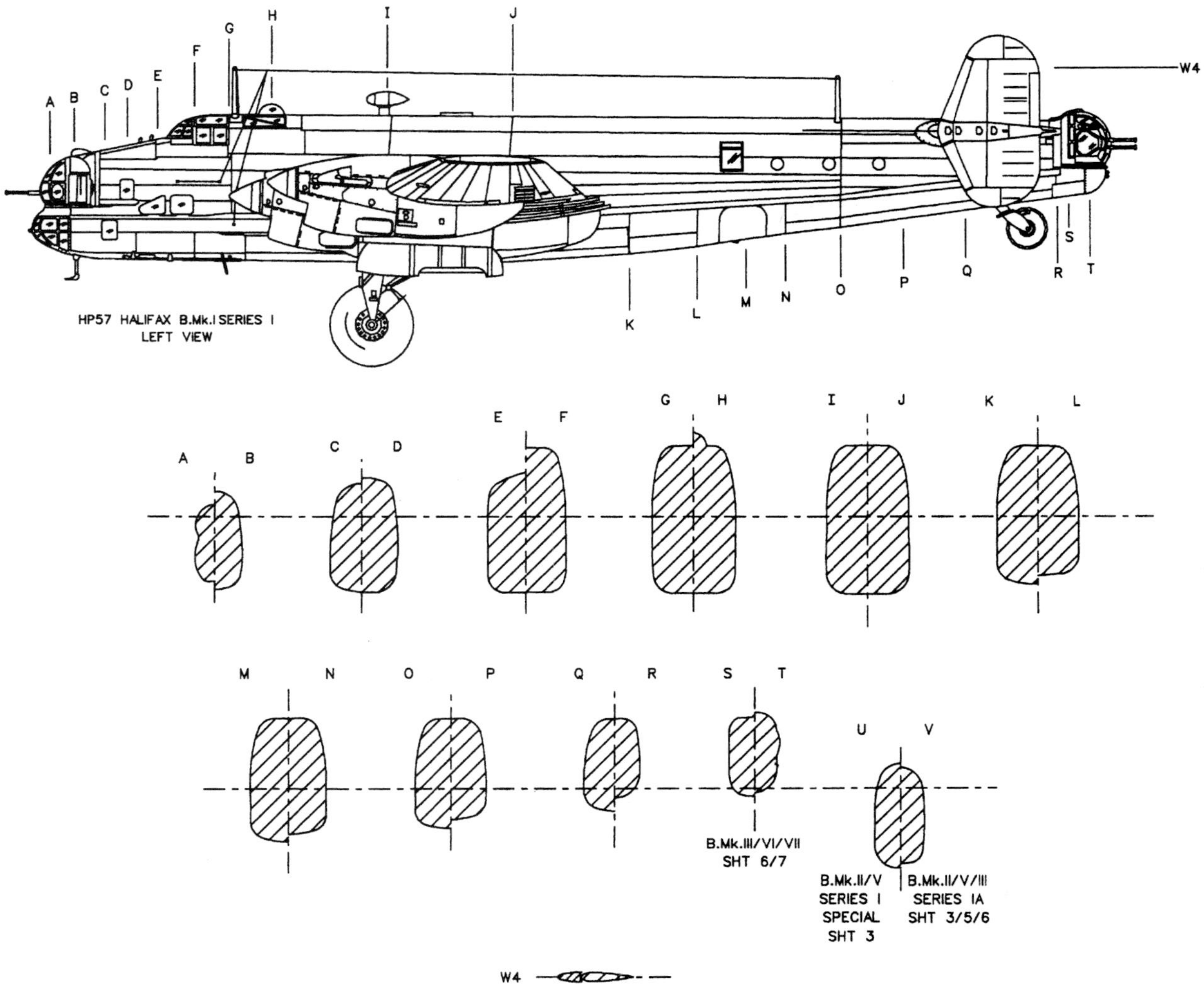

30

Top view of Handley Page Halifax B.Mk.I Series I *(Andrew Tattersall)*

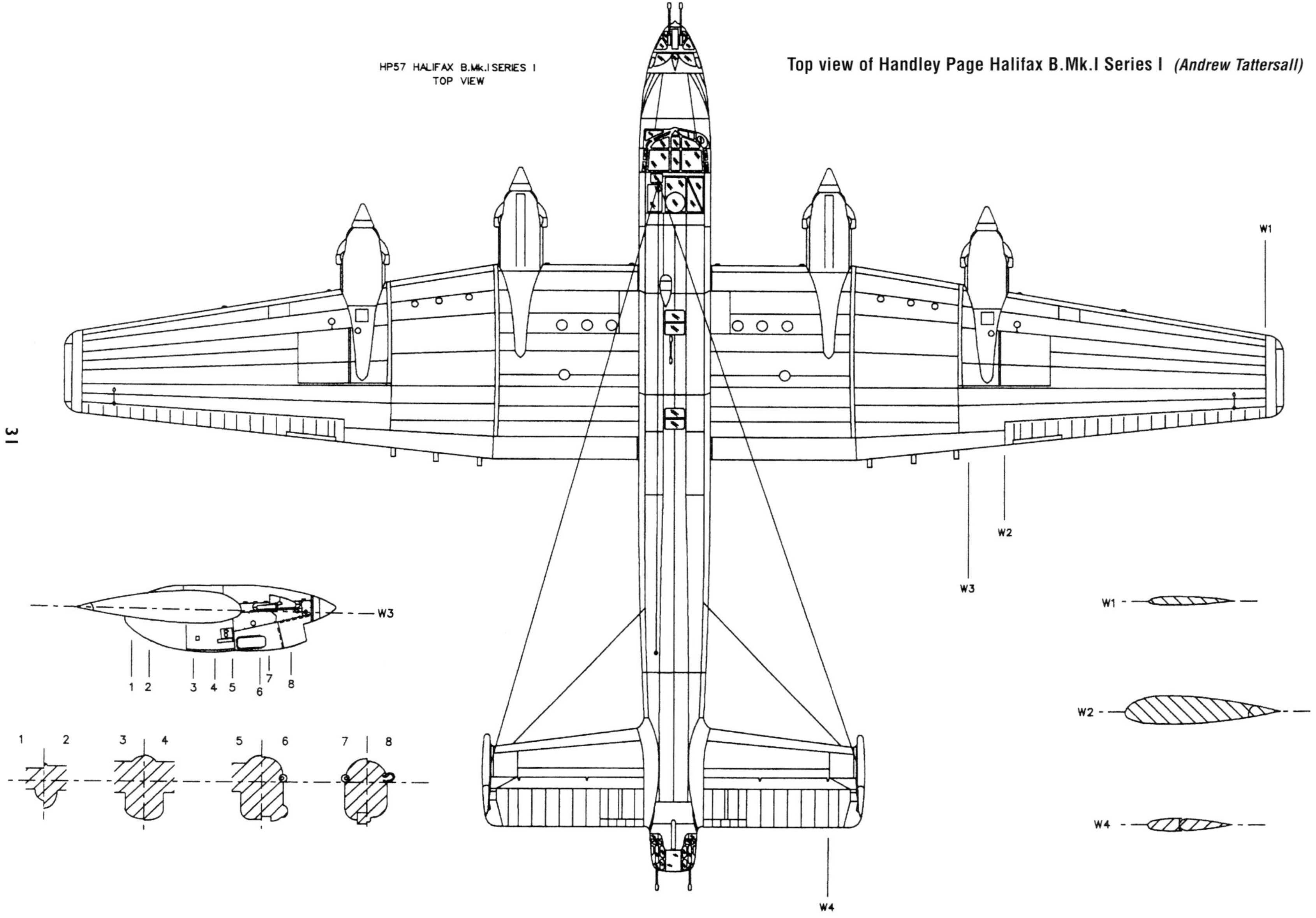

Bottom view of Handley Page Halifax B.Mk.I Series I *(Andrew Tattersall)*

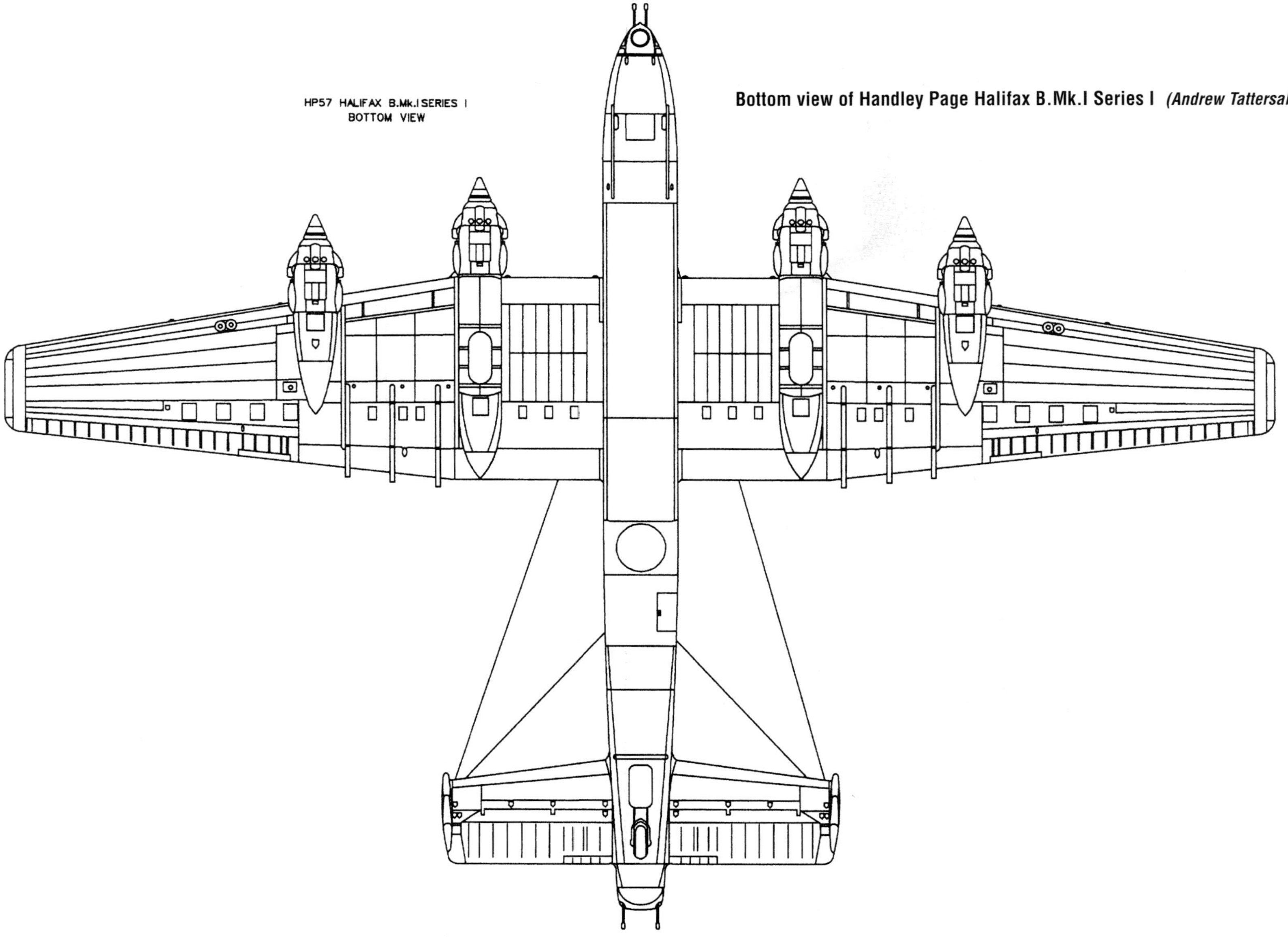

HP57 HALIFAX B.Mk.I SERIES I
BOTTOM VIEW

CHAPTER THREE

Handley Page HP.59 Halifax B.Mk.II Series I heavy bomber. This aircraft served with No. 408(B) Squadron RCAF during its early operations with Halifax aircraft. *(CF PL-116970)*

Handley Page HP.59 Halifax B.Mk.II Series I

Description: All-metal, mid wing, cantilever monoplane heavy bomber, equipped with Messier undercarriage and fixed tailwheel.

Powerplant: Four Rolls Royce Merlin XX V-12 cylinder engines rated at 1,390 h.p. with Delaney Gallay coolant radiators, driving three-blade Rotol constant speed compressed wood propellers

Dimensions:
- Height: 20 ft. 9 in. (6.32 m) (tail down, including DF loop antenna)
- Length: 70 ft. 1 in. (21.36 m)
- Wingspan: 98 ft. 8 in. (30.07 m)
- Wing Area: 1,250 sq. ft. (116.13 m2)
- Tailplane Span: 30 ft. 4 in. (9.25 m)
- Undercarriage Track: 24 ft. 8 in. (7.52 m)
- Dihedral Inner Mainplane: 00
- Dihedral Outer Mainplane: 20

Weights: Empty weight: 35,800 lbs. (16,238.9 kg.)
Max. takeoff weight: 60,000 lbs. (27,216 kg.)

Performance: Max speed: 254 mph at 12,750 ft. at 60,000 lbs. all up weight
(408.8 kph at 3,886.2 m at 27,216 kg. all up weight)

Economical Cruising Speed: 190 mph at 15,000 ft. (313.9 kph at 4572 m)

Takeoff Distance: 3,060 ft. at 60,000 lbs. to clear 50 ft. obstacle.
(932.7 m at 22,680 kg to clear 15.24 m obstacle)

Landing Distance: 2,550 ft. (777.3 m)

Initial rate of climb: 750 ft/min. at 60,000 lbs. all up weight
(228.6 m/min at 26,308.8 kg. all up weight)

Handley Page Halifax

Service Ceiling: 22,000 ft. (6,705.6 m)

Normal Range: 1900 miles with full tanks and 5,800 lbs. bomb load
(3,058 km. With full tanks and 2,631 kg. bomb load)

Armament: Boulton Paul Type C Nose Turret with two Browning Mk.II 0.303 machine guns and 1,000 rounds of ammunition per gun.

Boulton Paul Type C Mid Upper Turret with two Browning Mk.II 0.303 machine guns and 1,000 rounds of ammunition per gun.

Boulton Paul Type E Tail Turret with four Browning Mk.II 0.303 machine guns and 2,500 rounds of ammunition per gun.

Bomb Load: *Maximum 13,000 lbs.*

One 500 lb. bomb in each of six wing bomb bays
Two 2,000lb. bombs + six 1,000 lb. bombs in fuselage bomb bay

Or:

One 500 lb. bomb in each of six wing bomb bays
Four 2,000 bombs in fuselage bomb bay

Or:

One 500 lb. bomb in each of six wing bomb bays
Two 1,500 lb. mines + six 500 lb. bombs in fuselage bomb bay

Or:

One 500 lb. bomb in each of six wing bomb bays
Nine 500 lb. bombs in fuselage bomb bay

Or:

One 500 lb. bomb in each of six wing bomb bays
One 8,000 lb. bomb in fuselage bomb bay

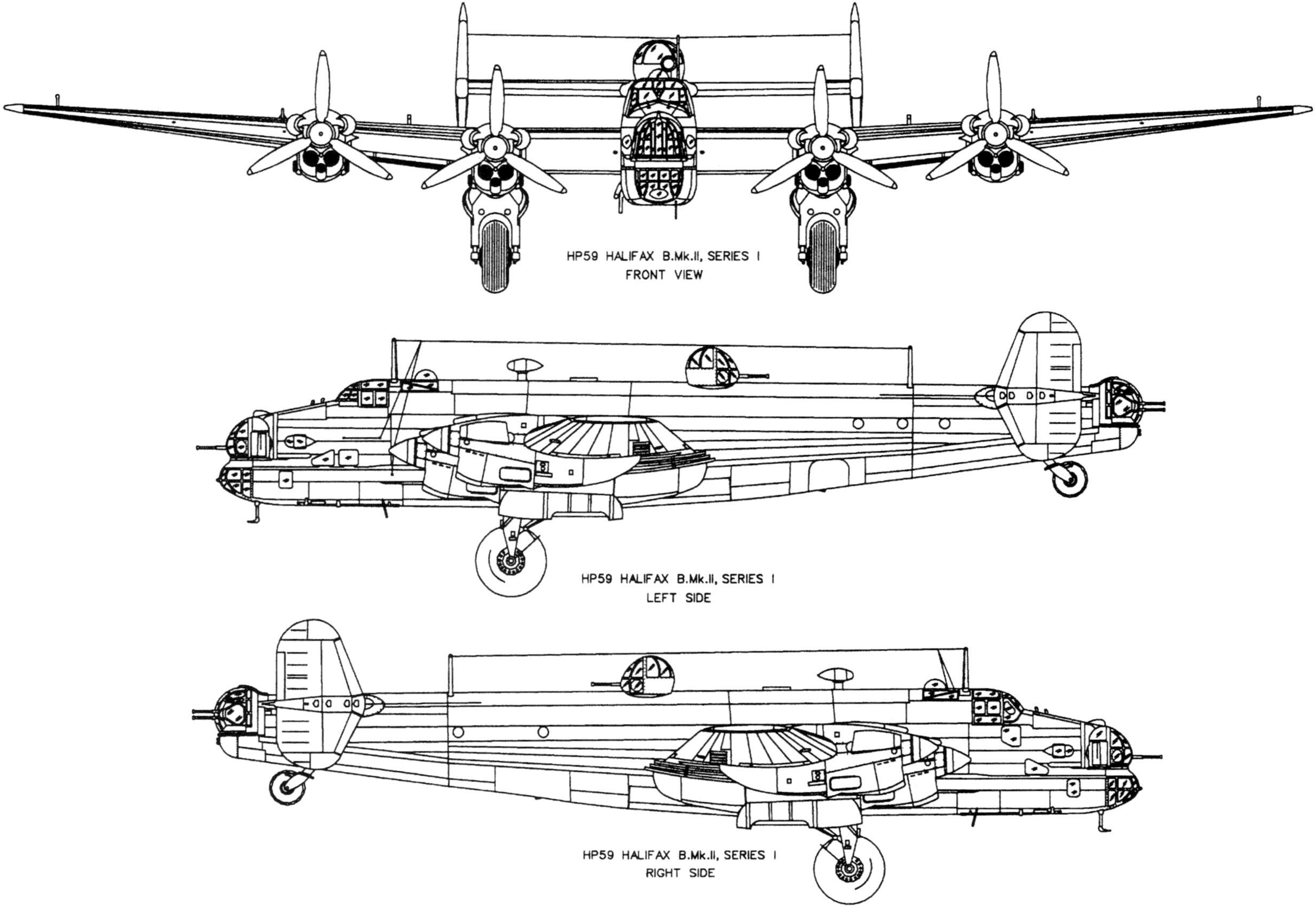

HP59 HALIFAX B.Mk.II, SERIES I
FRONT VIEW

HP59 HALIFAX B.Mk.II, SERIES I
LEFT SIDE

HP59 HALIFAX B.Mk.II, SERIES I
RIGHT SIDE

Front, left and right side views of Handley Page Halifax B.Mk.II Series I *(Andrew Tattersall)*

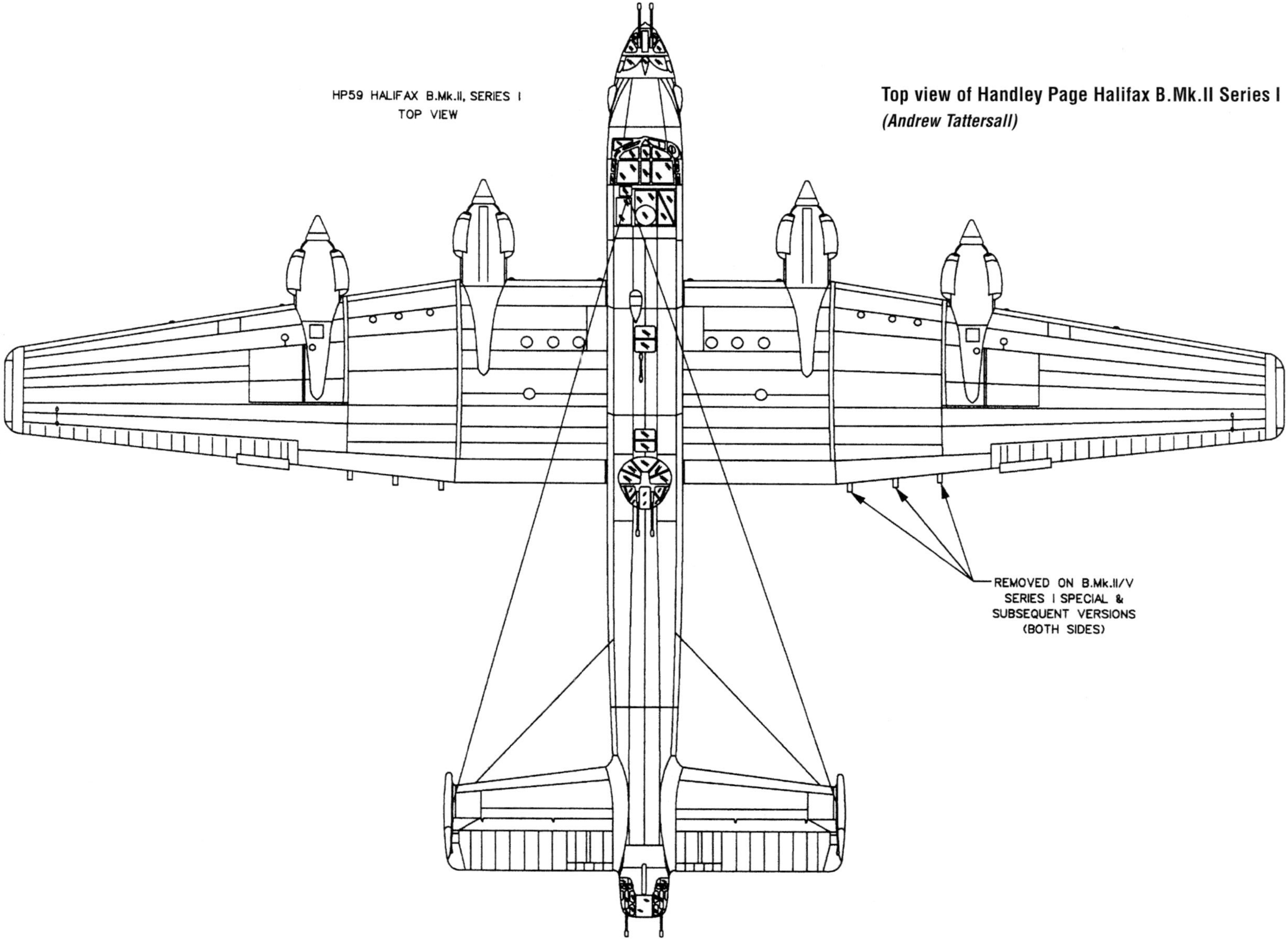

Top view of Handley Page Halifax B.Mk.II Series I
(Andrew Tattersall)

REMOVED ON B.Mk.II/V SERIES I SPECIAL & SUBSEQUENT VERSIONS (BOTH SIDES)

HP59 HALIFAX B.Mk.II, SERIES I
TOP VIEW

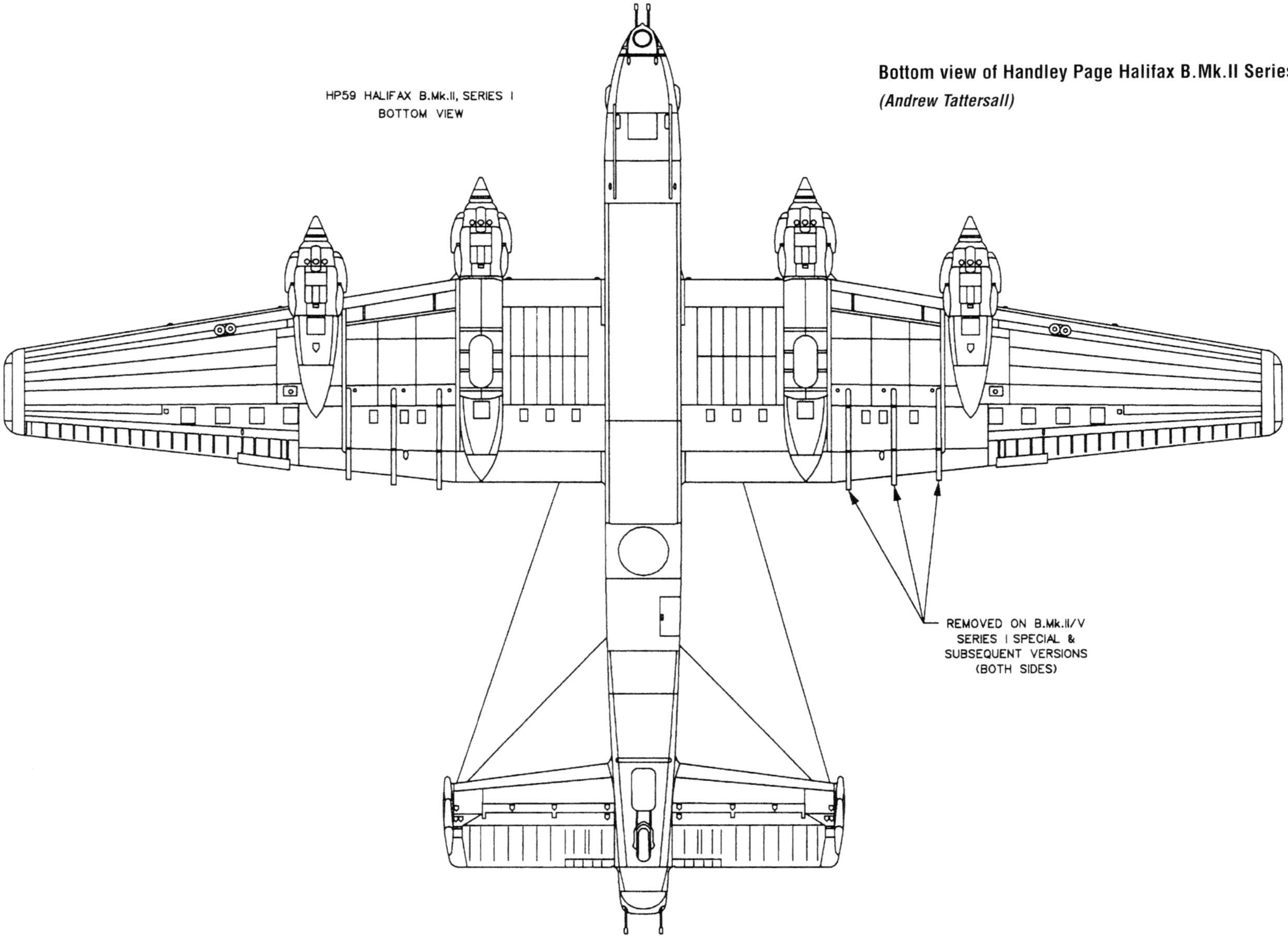

Bottom view of Handley Page Halifax B.Mk.II Series I
(Andrew Tattersall)

REMOVED ON B.Mk.II/V
SERIES I SPECIAL &
SUBSEQUENT VERSIONS
(BOTH SIDES)

HP59 HALIFAX B.Mk.II, SERIES I
BOTTOM VIEW

Handley Page Halifax

Handley Page HP.59 Halifax B.Mk.II Series I (Special), ZA*B of No. 10 Sqn RCAF, showing the "Z" nose fairing to advantage. *(CF PL-6899)*

Handley Page HP.59 Halifax B.Mk.II Series I (Special)

Description: All-metal, mid wing, cantilever monoplane heavy bomber, equipped with Messier undercarriage and fixed tailwheel.

Powerplant: Four Rolls Royce Merlin XX V-12 cylinder engines rated at 1,390 hp with Delaney Gallay coolant radiators, driving three-blade Rotol constant speed compressed wood propellers.

Dimensions:
- Height: 20 ft. 9 in. (6.32 m) (tail down, including DF loop antenna)
- Length: 70 ft. 1 in. (21.36 m)
- Wingspan: 98 ft. 8 in. (30.07 m)
- Wing Area: 1,250 sq. ft. (116.13 m2)
- Tailplane Span: 30 ft. 4 in. (9.25 m)
- Undercarriage Track: 24 ft. 8 in. (7.52 m)
- Dihedral Inner Mainplane: 00
- Dihedral Outer Mainplane: 20

Weights: Empty weight: 36,000 lbs. (16,329.6 kg.)
Max. takeoff weight: 61,500 lbs. (27,896.4 kg.)

Performance: Max speed: 254 mph at 12,750 ft. at 60,000 lbs. all up weight
(408.8 kph at 3,886.2 m at 27,216 kg. all up weight)

Economical Cruising Speed: 190 mph at 15,000 ft. (313.9 kph at 4572 m)

Takeoff distance: 3,060 ft. at 60,000 lbs. to clear 50 ft. obstacle
(932.7 m at 22,680 kg to clear 15.24 m obstacle)

CHAPTER THREE

Landing Distance: 2,550 ft. (777.3 m)

Initial rate of climb: 750 ft/min. at 60,000 lbs. all up weight
(228.6 m/min at 26,308.8 kg. all up weight)

Service Ceiling: 22,000 ft. (6,705.6 m)

Normal Range: 1900 miles with full tanks and 5,800 lb. bomb load
(3,058 km. With full tanks and 2,631 kg. bomb load)

Armament: Boulton Paul Type E Tail Turret with four Browning Mk.II 0.303 machine guns and 2,500 rounds of ammunition per gun.

Boulton Paul Type C Mid Upper Turret with two Browning Mk.II 0.303 machine guns and 1,000 rounds of ammunition per gun (some aircraft) Or,

Boulton Paul Type A Mk.VIII Mid Upper Turret with two Browning Mk.II 0.303 machine guns and 600 rounds of ammunition per gun (some aircraft)

Bomb Load: *Maximum 13,000 lbs.*

One 500 lb. bomb in each of six wing bomb bays
Two 2,000 lb. bombs + six 1,000 lb. bombs in fuselage bomb bay

Or:

One 500 lb. bomb in each of six wing bomb bays
Four 2,000 bombs in fuselage bomb bay

Or:

One 500 lb. bomb in each of six wing bomb bays
Two 1,500 lb. mines + six 500 lb. bombs in fuselage bomb bay

Or:

One 500 lb. bomb in each of six wing bomb bays
Nine 500 lb. bombs in fuselage bomb bay

Or:

One 500 lb. bomb in each of six wing bomb bays
One 8,000 lb. bomb in fuselage bomb bay

Front view of HP.59 Halifax B.Mk.II Series I (Special). *(CF PL-30839)*

Handley Page Halifax

Handley Page HP.59 Halifax B.Mk.IA, EY*B of 78 Sqn RAF, with early swept vertical stabilizer, showing the definitive glazed nose perspex. *(CF PL-11961)*

CHAPTER THREE

Handley Page HP.59 Halifax B.Mk.II Series Ia

Description: All-metal, mid wing, cantilever monoplane heavy bomber, equipped with Messier undercarriage and fixed tailwheel.

Powerplant: Four Rolls Royce Merlin 22 V-12 cylinder engines rated at 1,390 h.p. with Morris block coolant radiators, driving three-blade Rotol constant speed compressed wood propellers.

Dimensions:
- Height: 20 ft. 9 in. (6.32 m) (tail down, including DF loop antenna)
- Length: 71 ft. 7 in. (21.82 m)
- Wingspan: 98 ft. 8 in. (30.07 m)
- Wing Area: 1,250 sq. ft. (116.13 m2)
- Tailplane Span: 30 ft. 4 in. (9.25 m)
- Undercarriage Track: 24 ft. 8 in. (7.52 m)
- Dihedral Inner Mainplane: 00
- Dihedral Outer Mainplane: 20

Weights: Empty weight: 35,577 lbs. (16,137.7 kg.)
Max. takeoff weight: 60,000 lbs. (27,216 kg.)

Performance: Max speed: 250 mph at 13,000 ft. at 60,000 lbs. all up weight
(402.3 kph at 3,962.4 m at 27,216 kg. all up weight

Cruising Speed: 205 mph at 20,000 ft. (329.9 kph at 6,096 m)

Takeoff distance: 3,060 ft. at 60,000 lbs. to clear 50 ft. obstacle.
(932.7 m at 22,680 kg to clear 15.24 m obstacle)

Initial rate of climb: 750 ft/min. at 60,000 lbs. all up weight
(228.6 m/min at 26,308.8 kg. all up weight)

Service Ceiling: 21,500 ft. (9,752.4 m)

Normal Range: 1,660 miles with full tanks and 5,800 lb. bomb load
(2,671.4 km with full tanks and 2,631 kg. bomb load)

Armament: Single hand-operated Vickers K 0.303 machine gun in nose position

Boulton Paul Type A Mk.VIII Mid Upper Turret with two Browning Mk.II 0.303 machine guns and 600 rounds of ammunition per gun.

Boulton Paul Type E Tail Turret with four Browning Mk.II 0.303 machine guns and 2,500 rounds of ammunition per gun.

Bomb Load: *Maximum 13,000 lbs.*

One 500 lb. bomb in each of six wing bomb bays
Two 2,000lb. bombs + six 1,000 lb. bombs in fuselage bomb bay

    Or:

One 500 lb. bomb in each of six wing bomb bays
Four 2,000 bombs in fuselage bomb bay

    Or:

One 500 lb. bomb in each of six wing bomb bays
Two 1,500 lb. mines + six 500 lb. bombs in fuselage bomb bay

    Or:

One 500 lb. bomb in each of six wing bomb bays
Nine 500 lb. bombs in fuselage bomb bay

    Or:

One 500 lb. bomb in each of six wing bomb bays
One 8,000 lb. bomb in fuselage bomb bay

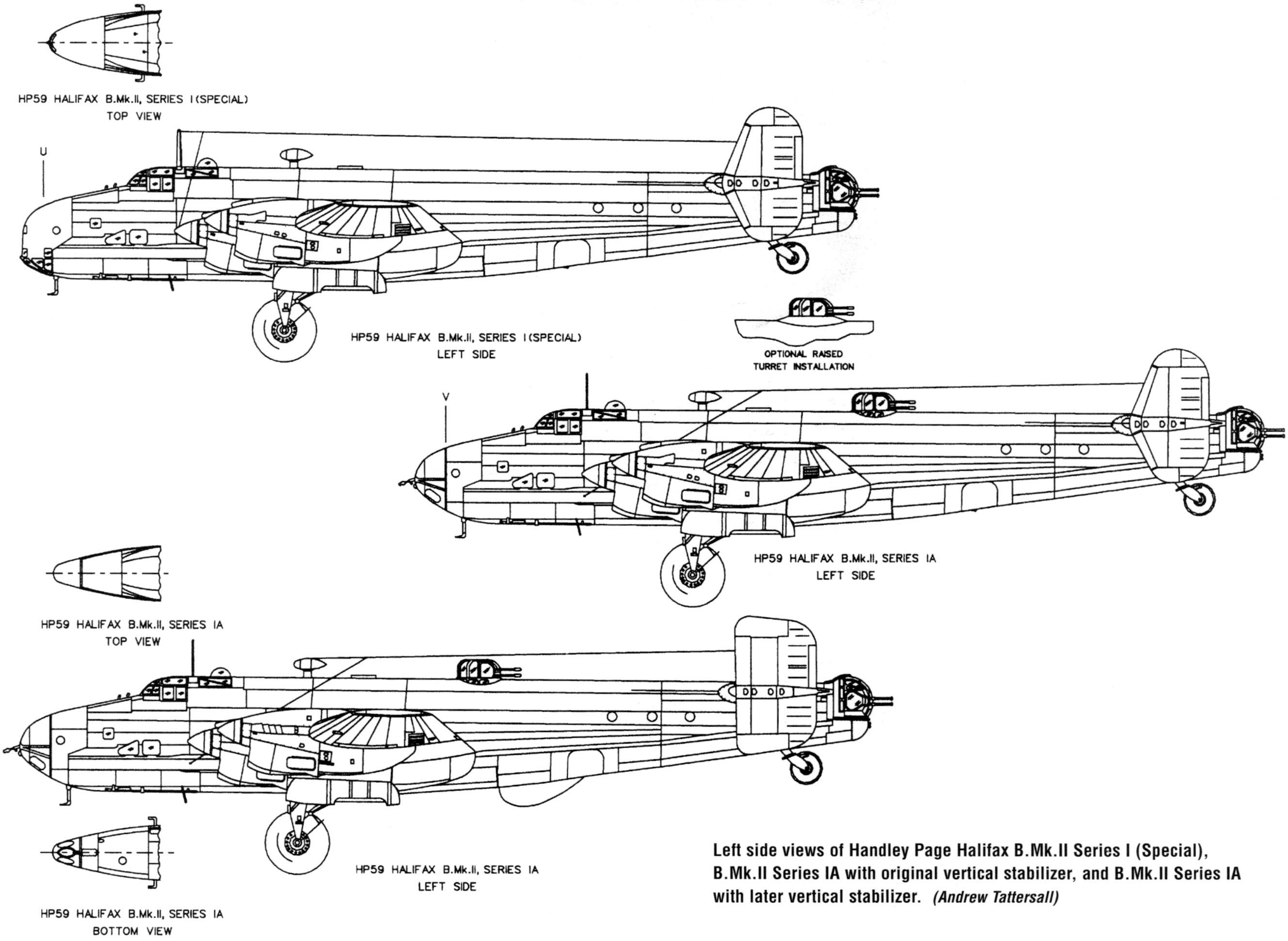

HP59 HALIFAX B.Mk.II, SERIES I (SPECIAL) TOP VIEW

HP59 HALIFAX B.Mk.II, SERIES I (SPECIAL) LEFT SIDE

OPTIONAL RAISED TURRET INSTALLATION

HP59 HALIFAX B.Mk.II, SERIES IA LEFT SIDE

HP59 HALIFAX B.Mk.II, SERIES IA TOP VIEW

HP59 HALIFAX B.Mk.II, SERIES IA LEFT SIDE

HP59 HALIFAX B.Mk.II, SERIES IA BOTTOM VIEW

Left side views of Handley Page Halifax B.Mk.II Series I (Special), B.Mk.II Series IA with original vertical stabilizer, and B.Mk.II Series IA with later vertical stabilizer. *(Andrew Tattersall)*

CHAPTER THREE

Handley Page HP.63 Halifax B.Mk.V Series I (Special) in service with 427(B) "Lion" Sqn RCAF undergoing maintenance at Leeming, Yorkshire, 16 November 1943. *(CF PL-26136)*

Handley Page HP.63 Halifax B.Mk.V Series I Special

Description: All-metal, mid wing, cantilever monoplane heavy bomber, equipped with Dowty lever suspension undercarriage and fixed tailwheel.

Powerplant: Four Rolls Royce Merlin XX V-12 cylinder engines rated at 1,390 h.p. with Delaney Gallay coolant radiators, driving three-blade Rotol constant speed compressed wood propellers.

Dimensions:
- Height: 20 ft. 9 in.(6.32 m) (tail down, including DF loop antenna
- Length: 70 ft. 1 in. (21.36 m)
- Wingspan: 98 ft. 8 in. (30.07 m)
- Wing Area: 1,250 sq. ft. (116.13 m2)
- Tailplane Span: 30 ft. 4 in. (9.25 m)
- Undercarriage Track: 24 ft. 8 in. (7.52 m)
- Dihedral Inner Mainplane: 00
- Dihedral Outer Mainplane: 20

Handley Page Halifax

Weights: Empty weight: 36,400 lbs. (16,511 kg.)
Max. takeoff weight: 61,500 lbs. (27,896.4 kg.)

Performance: Max speed: 254 mph at 12,750 ft. at 60,000 lbs. all up weight
(408.8 kph at 5783.4 m at 27,216 kg.)

Cruising Speed: 190 mph at 15,000 ft.

Takeoff distance: 3,060 ft. at 60,000 lbs. to clear 50 ft. obstacle.
(932.7 m at 22,680 kg to clear 15.24 m obstacle)

Initial rate of climb: 750 ft/min. at 60,000 lbs. all up weight
(228.6 m/min at 26,308.8 kg. all up weight)

Service Ceiling: 22,000 ft. (6,705.6 m)

Normal Range: 1900 miles with full tanks and 5,800 lb. bomb load
(3057.7 km with full tanks and 2,630.9 kg. bomb load)

Armament: Boulton Paul Type E Tail Turret with four Browning Mk.II 0.303 machine guns and 2,500 rounds of ammunition per gun.

Boulton Paul Type C Mid Upper Turret with two Browning Mk.II 0.303 machine guns and 1,000 rounds of ammunition per gun (some aircraft)

Boulton Paul Type A Mk.VIII Mid Upper Turret with two Browning Mk.II 0.303 machine guns and 600 rounds of ammunition per gun (some aircraft)

Bomb Load: *Maximum 13,000 lbs.*

One 500 lb. bomb in each of six wing bomb bays
Two 2,000lb. bombs + six 1,000 lb. bombs in fuselage bomb bay

Or:

One 500 lb. bomb in each of six wing bomb bays
Four 2,000 bombs in fuselage bomb bay

Or:

One 500 lb. bomb in each of six wing bomb bays
Two 1,500 lb. mines + six 500 lb. bombs in fuselage bomb bay

Or:

One 500 lb. bomb in each of six wing bomb bays
Nine 500 lb. bombs in fuselage bomb bay

Or:

One 500 lb. bomb in each of six wing bomb bays
One 8,000 lb. bomb in fuselage bomb bay

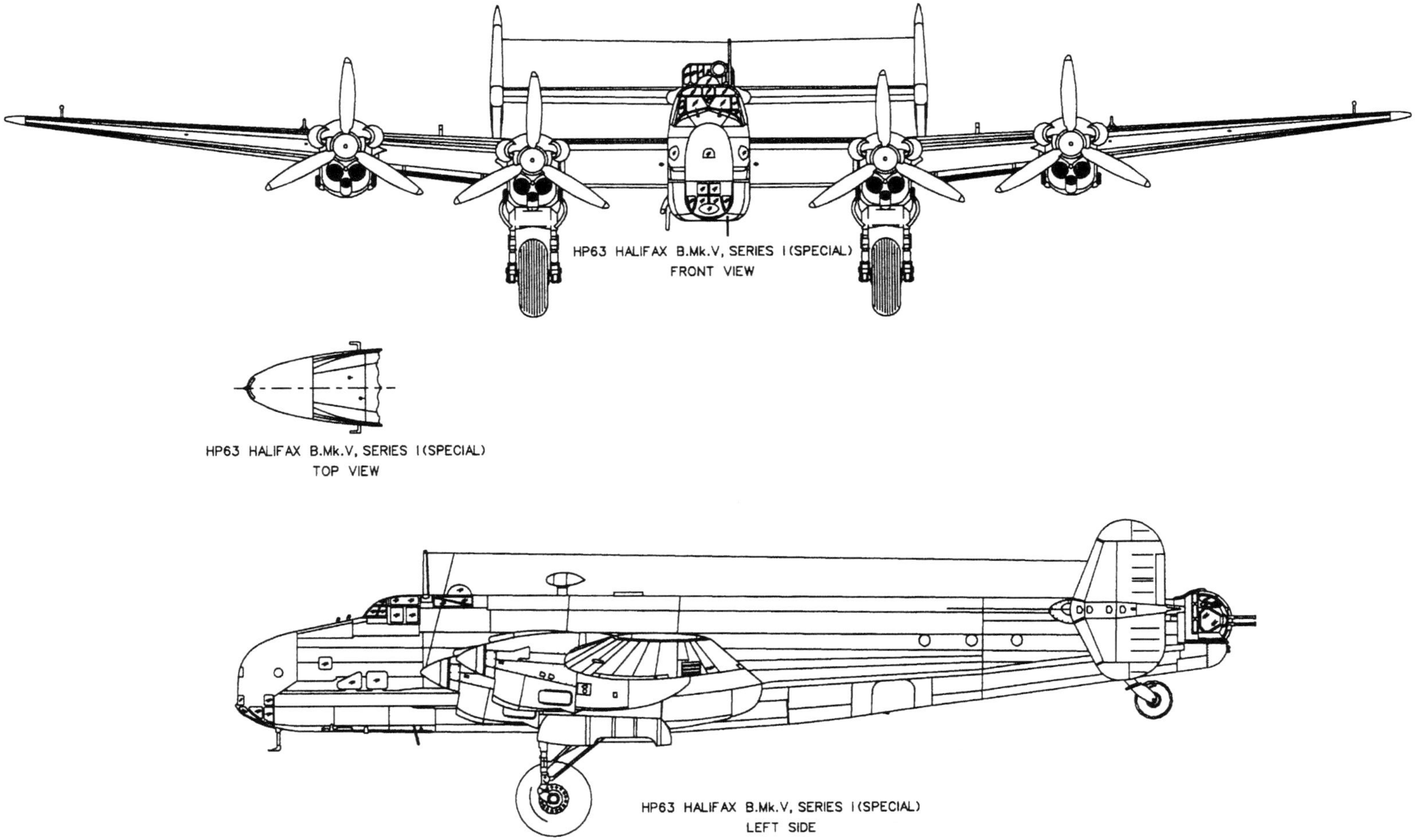

Front and left side view of Handley Page HP.63 Halifax B.Mk.V Series I (Special) *(Andrew Tattersall)*

Handley Page Halifax

Handley Page HP.63 Halifax B.Mk.V Series Ia in service with a Heavy Conversion Unit of the RCAF taxying out for a training sortie, 20 July 1944. *(CF PL-31506)*

CHAPTER THREE

Handley Page HP.63 Halifax B.Mk.V Series Ia

Description: All-metal, mid wing, cantilever monoplane heavy bomber, equipped with Dowty lever suspension undercarriage and fixed tailwheel.

Powerplant: Four Rolls Royce Merlin 22 V-12 cylinder engines rated at 1,390 hp with Morris block coolant radiators, driving three-blade Rotol constant speed compressed wood propellers

Dimensions:
- Height: 20 ft. 9 in. (6.32 m) (tail down, including DF loop antenna)
- Length: 71 ft. 7 in. (21.82 m)
- Wingspan: 98 ft. 8 in. (30.07 m)
- Wing Area: 1,250 sq. ft. (116.13 m2)
- Tailplane Span: 30 ft. 4 in. (9.25 m)
- Undercarriage Track: 24 ft. 8 in. (7.52 m)
- Dihedral Inner Mainplane: 00
- Dihedral Outer Mainplane: 20

Weights: Empty weight: 35,577 lbs. (16,137.7 kg.)
Max. takeoff weight: 60,000 lbs. (27,216 kg.)

Performance: Max speed: 250 mph at 13,000 ft. at 60,000 lbs. all up weight
(402.3 kph at 3.962.4 m at 27,216 kg. all up weight)

Cruising Speed: 205 mph at 20,000 ft. (329.9 kph at 6,096 m)

Takeoff distance: 3,060 ft. at 60,000 lbs. to clear 50 ft. obstacle.
(932.7 m at 22,680 kg to clear 15.24 m obstacle)

Initial rate of climb: 750 ft/min. at 60,000 lbs. all up weight
(228.6 m/min at 26,308.8 kg. all up weight)

Service Ceiling: 22,000 ft. (6,705.6 m)

Normal Range: 1,660 miles with full tanks and 5,800 lb. bomb load
(2,671.4 km with full tanks and 2,630.9 kg. bomb load)

Armament: Single hand-operated Vickers K 0.303 machine gun in nose position

Boulton Paul Type A Mk.VIII Mid Upper Turret with two Browning Mk.II 0.303 machine guns and 600 rounds of ammunition per gun.

Boulton Paul Type E Tail Turret with four Browning Mk.II 0.303 machine guns and 2,500 rounds of ammunition per gun.

Bomb Load: *Maximum 13,000 lbs.*

One 500 lb. bomb in each of six wing bomb bays
Two 2,000lb. bombs + six 1,000 lb. bombs in fuselage bomb bay

Or:

One 500 lb. bomb in each of six wing bomb bays
Four 2,000 bombs in fuselage bay

Or:

One 500 lb. bomb in each of six wing bomb bays
Two 1,500 lb. mines + six 500 lb. bombs in fuselage bomb bay

Or:

One 500 lb. bomb in each of six wing bomb bays
Nine 500 lb. bombs in fuselage bomb bay

Or:

One 500 lb. bomb in each of six wing bomb bays
One 8,000 lb. bomb in fuselage bomb bay

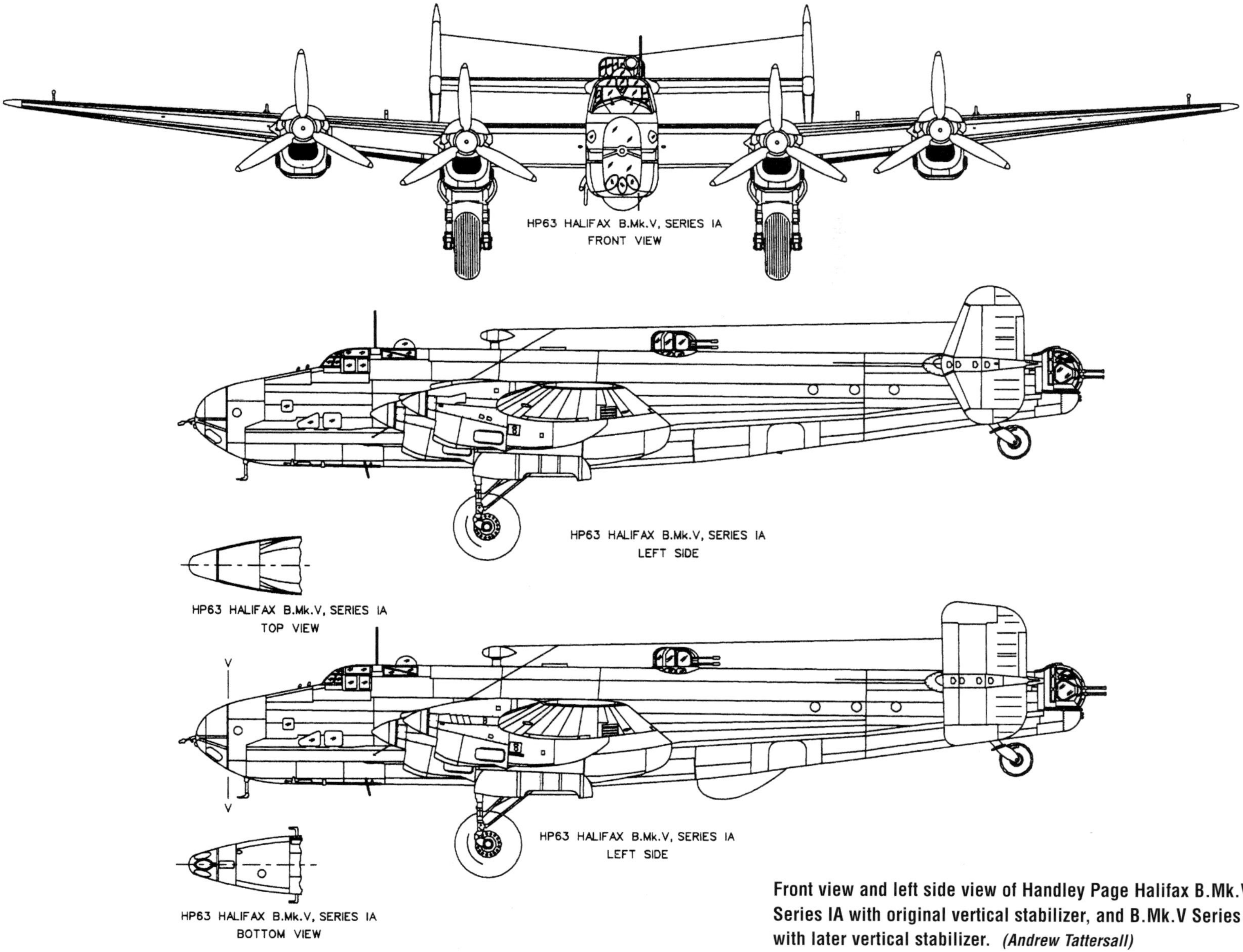

HP63 HALIFAX B.Mk.V, SERIES IA
FRONT VIEW

HP63 HALIFAX B.Mk.V, SERIES IA
LEFT SIDE

HP63 HALIFAX B.Mk.V, SERIES IA
TOP VIEW

HP63 HALIFAX B.Mk.V, SERIES IA
LEFT SIDE

HP63 HALIFAX B.Mk.V, SERIES IA
BOTTOM VIEW

Front view and left side view of Handley Page Halifax B.Mk.V Series IA with original vertical stabilizer, and B.Mk.V Series IA with later vertical stabilizer. *(Andrew Tattersall)*

CHAPTER THREE

Prototype of Handley Page HP.61 B.Mk.III, converted from Handley Page HP.59 B.MK.II Series I (Special).
(CF PL-116966)

VARIANTS POWERED BY BRISTOL HERCULES RADIAL ENGINES

Handley Page HP.61 Halifax B.Mk.III

Description: All-metal, mid wing, cantilever monoplane heavy bomber, equipped with Messier undercarriage and retractable tailwheel.

Powerplant: Four Bristol Hercules XVI 18 cylinder radial engines rated at 1,615 hp driving three-blade DeHavilland Hydromatic metal propellers.

Dimensions:
- Height: 20 ft. 9 in. (6.32 m) (tail down, including DF loop antenna)
- Length: 71 ft. 7 in. (21.82 m)
- Wingspan: 98 ft. 8 in. (30.07 m) or 104 ft. 2 in. (31.76 m) with extended wingtips
- Wing Area: 1,250 sq. ft. (116.13 m2) or 1,275 sq. ft. (118.45 m2)(with extended wingtips
- Tailplane Span: 30 ft. 4 in. (9.25 m)
- Undercarriage Track: 24 ft. 8 in. (7.52 m)
- Dihedral Inner Mainplane: 00
- Dihedral Outer Mainplane: 20

Weights: Empty weight: 38,240 lbs. (17,345.7 kg.)
Max. takeoff weight: 68,000 lbs. (30,844.8 kg,)

Performance: Max speed: 320 mph at 22,000 ft. at 65,000 lbs. all up weight
(515 kph at 6,705.6 m at 29,484 kg. all up weight)

Cruising Speed: 272 mph at 20,000 ft. (437.7 kph at 6096 m)

Stalling Speed: 92 mph (148.1 kph)

Takeoff distance: 3,705 ft. at 65,000 lbs. to clear 50 ft. obstacle.
(1129.3 m at 29,484 kg. To clear 15.24 m. obstacle)

Initial rate of climb: 750 ft/min. at 60,000 lbs. all up weight
(228.6 m/min at 18,288 kg. all up weight)

Service Ceiling: 24,000 ft. (7,315.2 m.)

Normal Range: 1,660 miles with full tanks and 5,800 lb. bomb load
(2,671.4 km. with full tanks and 2,630.9 kg. bomb load)

Max. Range: 3,220 miles at 195 mph
(5,181.9 km. at 313.8 kph)

Handley Page Halifax

Armament: Single hand-operated Vickers K 0.303 machine gun in nose position

Boulton Paul Type A Mk.VIII Mid Upper Turret with two Browning Mk.II 0.303 machine guns and 600 rounds of ammunition per gun.

Boulton Paul Type E Tail Turret with four Browning Mk.II 0.303 machine guns and 2,500 rounds of ammunition per gun.

Single 0.50 in. Browning machine gun in ventral position (some aircraft)

Bomb Load: *Maximum 13,000 lbs.*

One 500 lb. bomb in each of six wing bomb bays
Two 2,000lb. bombs + six 1,000 lb. bombs in fuselage bomb bay

   Or:

One 500 lb. bomb in each of six wing bomb bays
Four 2,000 bombs in fuselage bomb bay

   Or:

One 500 lb. bomb in each of six wing bomb bays
Two 1,500 lb. mines + six 500 lb. bombs in fuselage bomb bay

   Or:

One 500 lb. bomb in each of six wing bomb bays
Nine 500 lb. bombs in fuselage bomb bay

   Or:

One 500 lb. bomb in each of six wing bomb bays
One 8,000 lb. bomb in fuselage bomb bay

Radar: H2S mounted in ventral position in place of 0.50 in. machine gun.

Handley Page HP.61 B.Mk.III, late production, with extended wingtips. *(CF PL-41623)*

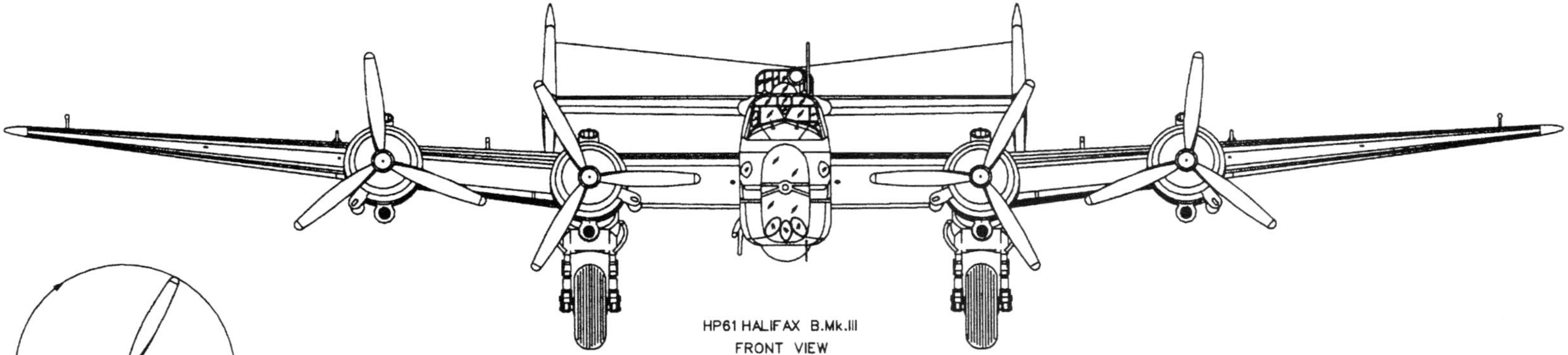

HP61 HALIFAX B.Mk.III
FRONT VIEW

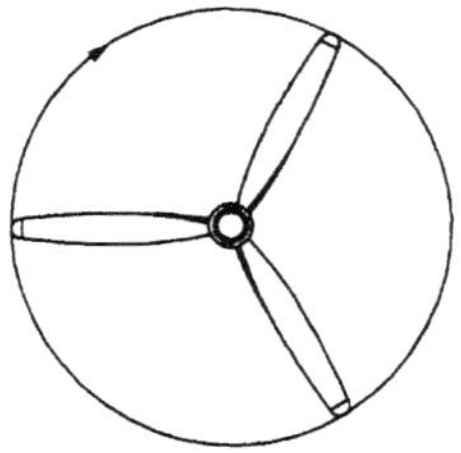

HP61 HALIFAX B.Mk.III,
LEFT SIDE

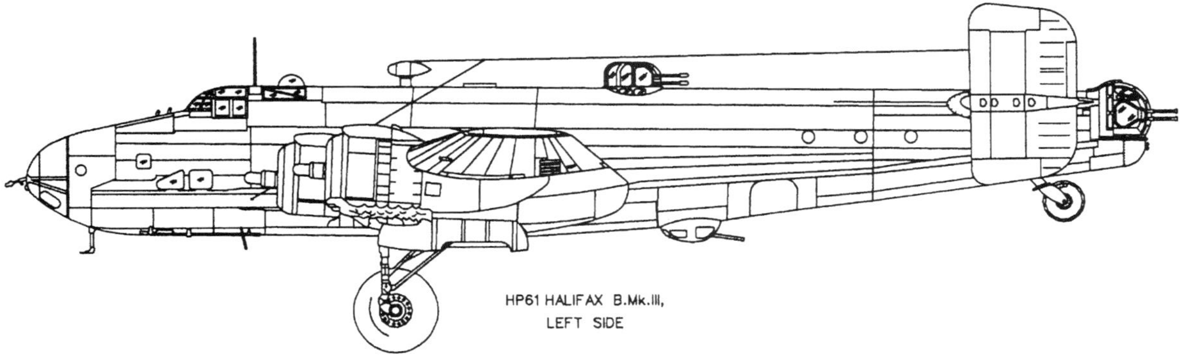

HP61 HALIFAX B.Mk.III,
LEFT SIDE

DE HAVILLAND HYDROMATIC: 13'0" DIA.
TYPE 55/18
Mks III/VI/VII

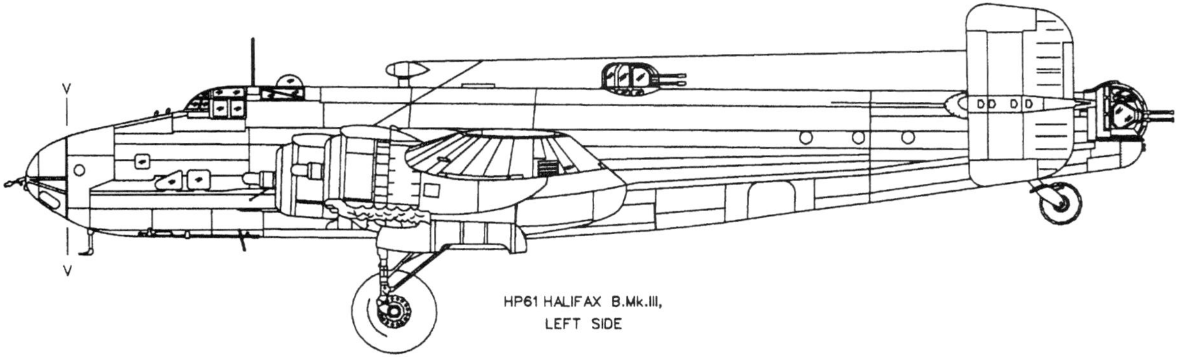

Front left side of an early production aircraft, and left side early production with underside .50 calibre gun turret, Handley Page HP.61 Halifax B.Mk.III.
(Andrew Tattersall)

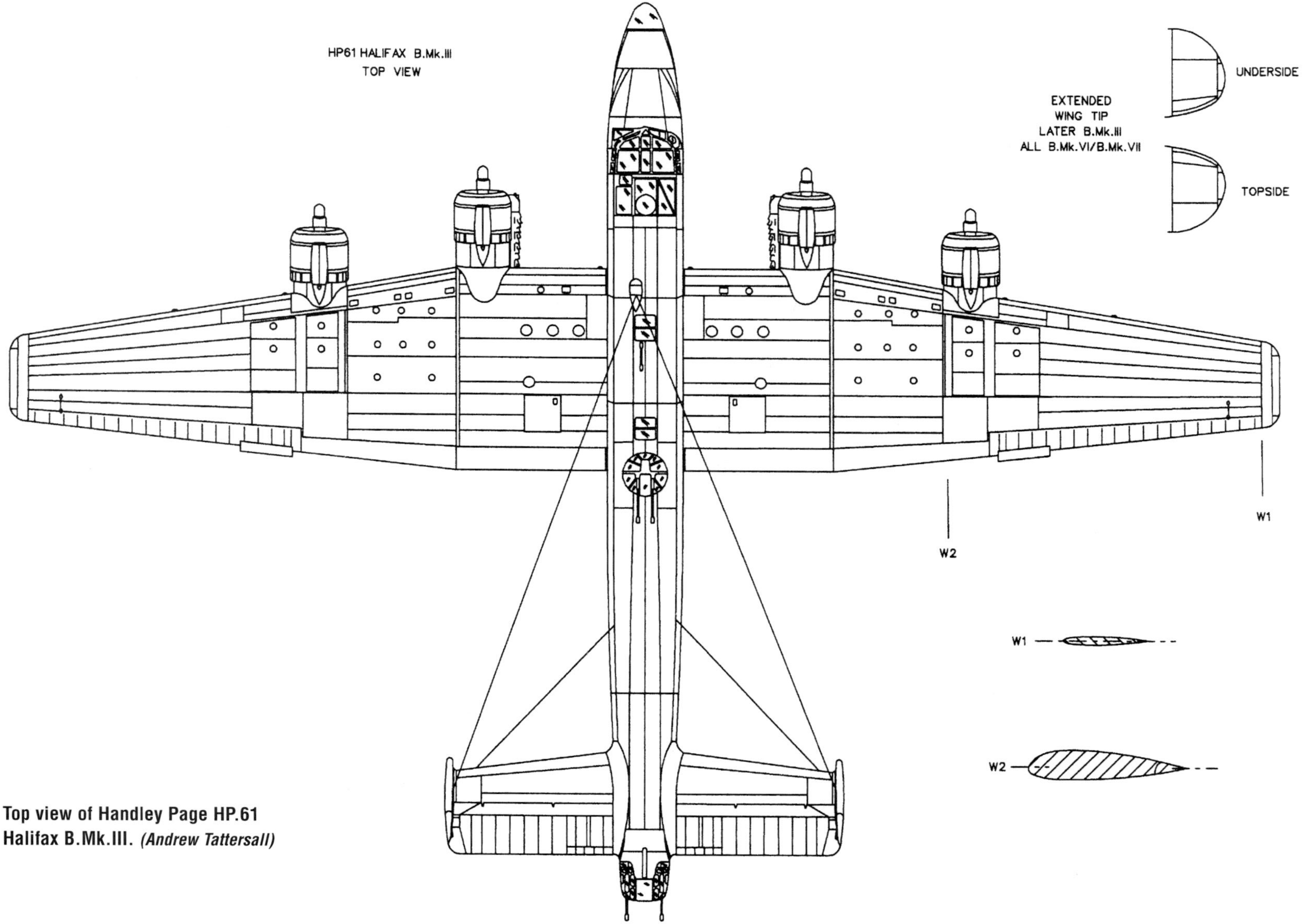

Top view of Handley Page HP.61 Halifax B.Mk.III. *(Andrew Tattersall)*

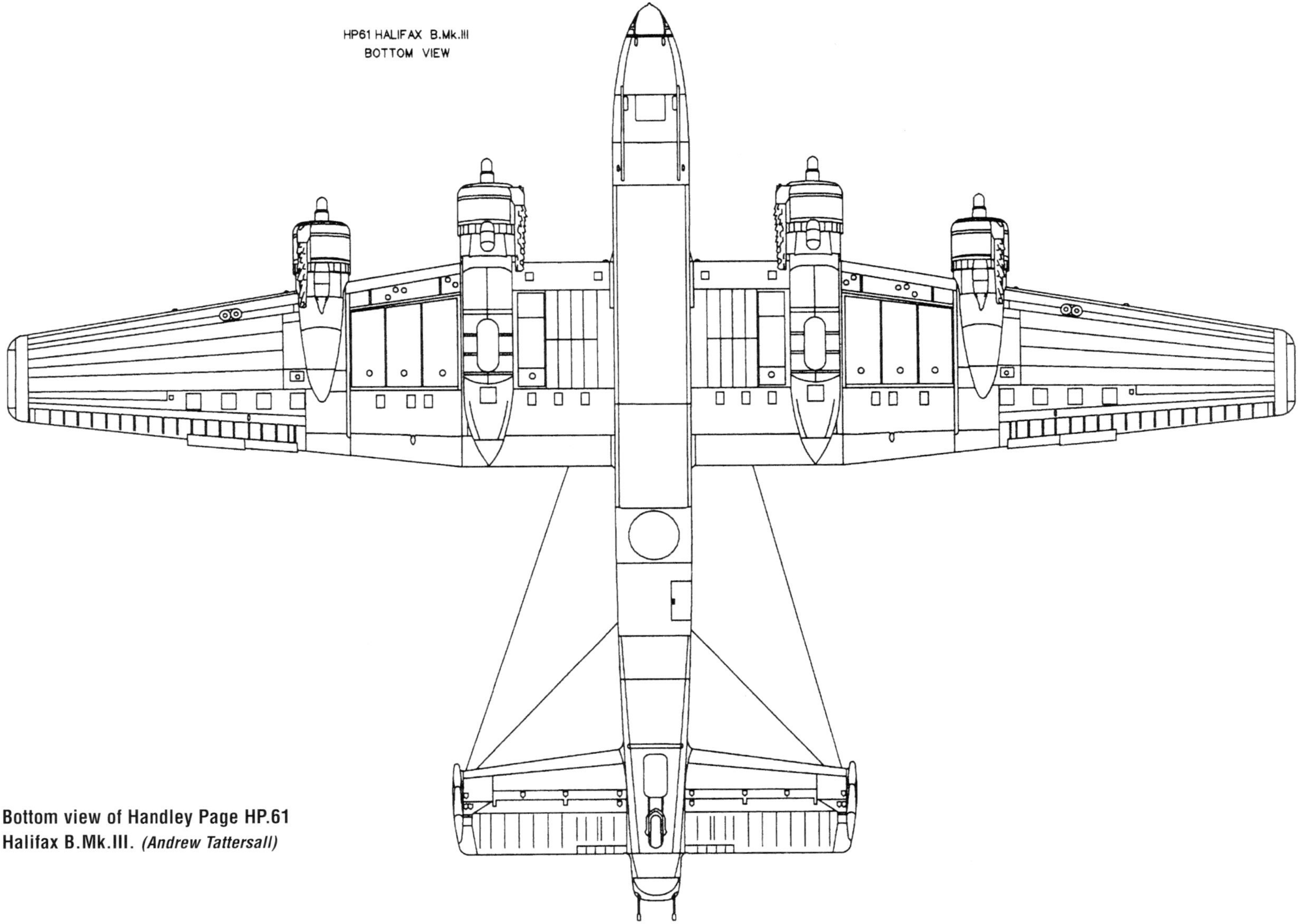

HP61 HALIFAX B.MK.III
BOTTOM VIEW

Bottom view of Handley Page HP.61 Halifax B.Mk.III. *(Andrew Tattersall)*

Handley Page Halifax

Halifax B.Mk.VI, serial number RS201, at the Winter Experimental Establishment, RCAF Station Namao (Edmonton), Alberta in the winter of 1946. *(CF PL-130106)*

Handley Page HP.61 Halifax B.Mk.VI

Description: All-metal, mid wing, cantilever monoplane heavy bomber, equipped with Messier undercarriage and retractable tailwheel.

Powerplant: Four Bristol Hercules 100 18-cylinder radial engines rated at 1,675 hp driving three-blade DeHavilland Hydromatic metal propellers.

Dimensions:
- Height: 20 ft. 9 in. (6.32 m) (tail down, including DF loop antenna)
- Wingspan: 104 ft. 2 in.
- Length: 71 ft. 7 in. (21.82 m)
- Wing Area: 1,275 sq. ft. (118.45 m2)
- Tailplane Span: 30 ft. 4 in. (9.25 m)
- Undercarriage Track: 24 ft. 8 in. (7.52 m)
- Dihedral Inner Mainplane: 00
- Dihedral Outer Mainplane: 20

CHAPTER THREE

Weights: Empty weight: 35,577 lbs. (16,137.7 kg.)
Normal all up weight: 65,000 lbs. (28,340 kg.)
Max. takeoff weight: 68,000 lbs. (30,844.8 kg,)

Performance: Max speed: 320 mph at 22,000 ft. at 65,000 lbs. all up weight
(515 kph at 6,705.6 m at 29,484 kg. all up weight)

Cruising Speed: 272 mph at 20,000 ft.
(437.7 kph at 6,096 m)

Stalling Speed: 92 mph (148.1 kph)

Takeoff distance: 3,705 ft. at 65,000 lbs. to clear 50 ft. obstacle.
(1,129.3 m at 28,340 kg. to clear 15.24 m obstacle)

Initial rate of climb: 750 ft/min. at 60,000 lbs. all up weight
(228.6 m/min at 27,216 kg. all up weight)

Service Ceiling: 24,000 ft. (7,315 m)

Normal Range: 1,660 miles with full tanks and 5,800 lb. bomb load
(2,671.4 km with full tanks and 2,630.9 kg. bomb load)

Max. Range: 3,220 miles at 195 mph at 10,000 ft.
(5,182 km at 313.8 kph at 3,048 m)

Armament: Single hand-operated Vickers K 0.303 machine gun in nose position

Boulton Paul Type A Mk.VIII Mid Upper Turret with two Browning Mk.II 0.303 machine guns and 600 rounds of ammunition per gun.

Boulton Paul Type E Tail Turret with four Browning Mk.II 0.303 machine guns and 2,500 rounds of ammunition per gun

Bomb Load: *Maximum 14,500 lbs.*

One 750 lb. bomb in each of six wing bomb bays
Two 2,000 lb. bombs + six 1,000 lb. bombs in fuselage bomb bay

Or:

One 750 lb. bomb in each of six wing bomb bays
Four 2,000 lb. bombs + six 500 lb. bombs in fuselage bomb bay

Or:

One 750 lb. bomb in each of six wing bomb bays
Two 1,500 lb. mines + six 500 lb. bombs in fuselage bomb bay

Or:

Nine 500 lb. bombs + two 4,000 lb. bombs in fuselage bomb bay

Or:

Nine 500 lb. bombs + one 8,000 lb. bomb in fuselage bomb bay

Radar: H2S permanently mounted in ventral position.

The HP.61 B.Mk.VI was equipped with a pressurized fuel system, additional fuel capacity, and sand filters in the carburettor intakes. This variant did not serve with RCAF Squadrons in 6 Group, Bomber Command, but one was tested at the WEE (Winter Experimental Establishment) at RCAF Station Namao (Edmonton) post war.

Handley Page Halifax

Handley Page HP.61 Halifax B.Mk.VII of 420(B) "Snowy Owl" Sqn taxiing for takeoff, 9 April 1945. The extended wingtips, upper and rear gun turrets and under fuselage radar installation are well displayed in this view. *(CF PL-43478)*

Handley Page HP.61 Halifax B.Mk.VII

Description: All-metal, mid wing, cantilever monoplane heavy bomber, equipped with Messier undercarriage and retractable tailwheel.

Powerplant: Four Bristol Hercules XVI 18-cylinder radial engines rated at 1,615 hp driving three-blade DeHavilland Hydromatic metal propellers.

Dimensions:
- Height: 20 ft. 9 in. (6.32 m) (tail down, including DF loop antenna)
- Wingspan: 104 ft. 2 in.
- Length: 71 ft. 7 in. (21.82 m)
- Wing Area: 1,275 sq. ft. (118.45 m2)
- Tailplane Span: 30 ft. 4 in. (9.25 m)
- Undercarriage Track: 24 ft. 8 in. (7.52 m)
- Dihedral Inner Mainplane: 00
- Dihedral Outer Mainplane: 20

CHAPTER THREE

Weights: Empty weight: 35,577 lbs. (16,137.7 kg.)
Normal all up weight: 65,000 lbs. (28,340 kg.)
Max. takeoff weight: 65,000 lbs. (28,340 kg.)

Performance: Max speed: 320 mph at 22,000 ft. at 65,000 lbs. all up weight
(515 kph at 6,705.6 m. at 29,484 kg. all up weight)

Cruising Speed: 272 mph at 20,000 ft.
(437.7 kph at 6,096 m)

Stalling Speed: 92 mph (148.1 kph)

Takeoff distance: 3,705 ft. at 65,000 lbs. to clear 50 ft. obstacle.
(1,129.3 m. at 29,484 kg. to clear 15.24 m obstacle)

Initial rate of climb: 750 ft/min. at 60,000 lbs. all up weight
(228.6 m/min at 27,216 kg. all up weight)

Service Ceiling: 24,000 ft. (8,179.2 m)

Normal Range: 1,660 miles with full tanks and 5,800 lb. bomb load
(2,671.4 km. with full tanks and 2,630.9 kg. bomb load)

Max. Range: 3,220 miles at 195 mph at 10,000 ft.
(5,182 km at 313.8 kph at 3,048 m)

Armament: Single hand-operated Vickers K 0.303 machine gun in nose position

Boulton Paul Type A Mk.VIII Mid Upper Turret with two Browning Mk.II 0.303 machine guns and 600 rounds of ammunition per gun.

Boulton Paul Type E Tail Turret with four Browning Mk.II 0.303 machine guns and 2,500 rounds of ammunition per gun.

Bomb Load: *Maximum 13,000 lbs.*

One 500 lb. bomb in each of six wing bomb bays
Two 2,000lb. bombs + six 1,000 lb. bombs in fuselage bomb bay

Or:

One 500 lb. bomb in each of six wing bomb bays
Four 2,000 bombs in fuselage bomb bay

Or:

One 500 lb. bomb in each of six wing bomb bays
Two 1,500 lb. mines + six 500 lb. bombs in fuselage bomb bay

Or:

One 500 lb. bomb in each of six wing bomb bays
Nine 500 lb. bombs in fuselage bomb bay

Or:

One 500 lb. bomb in each of six wing bomb bays
One 8,000 lb. bomb in fuselage bomb bay

Radar: H2S mounted in ventral position.

The HP.61 Halifax B.Mk.VII was similar to the B.Mk.VI, but was equipped with Bristol Hercules XVI engines, as Halifax airframe production had exceeded the supply of Bristol Hercules 100 engines. The majority of B.Mk.VII aircraft served with 408(B), 415(B), 420(B), 426(B), and 432(B) Squadrons of 6 (RCAF)

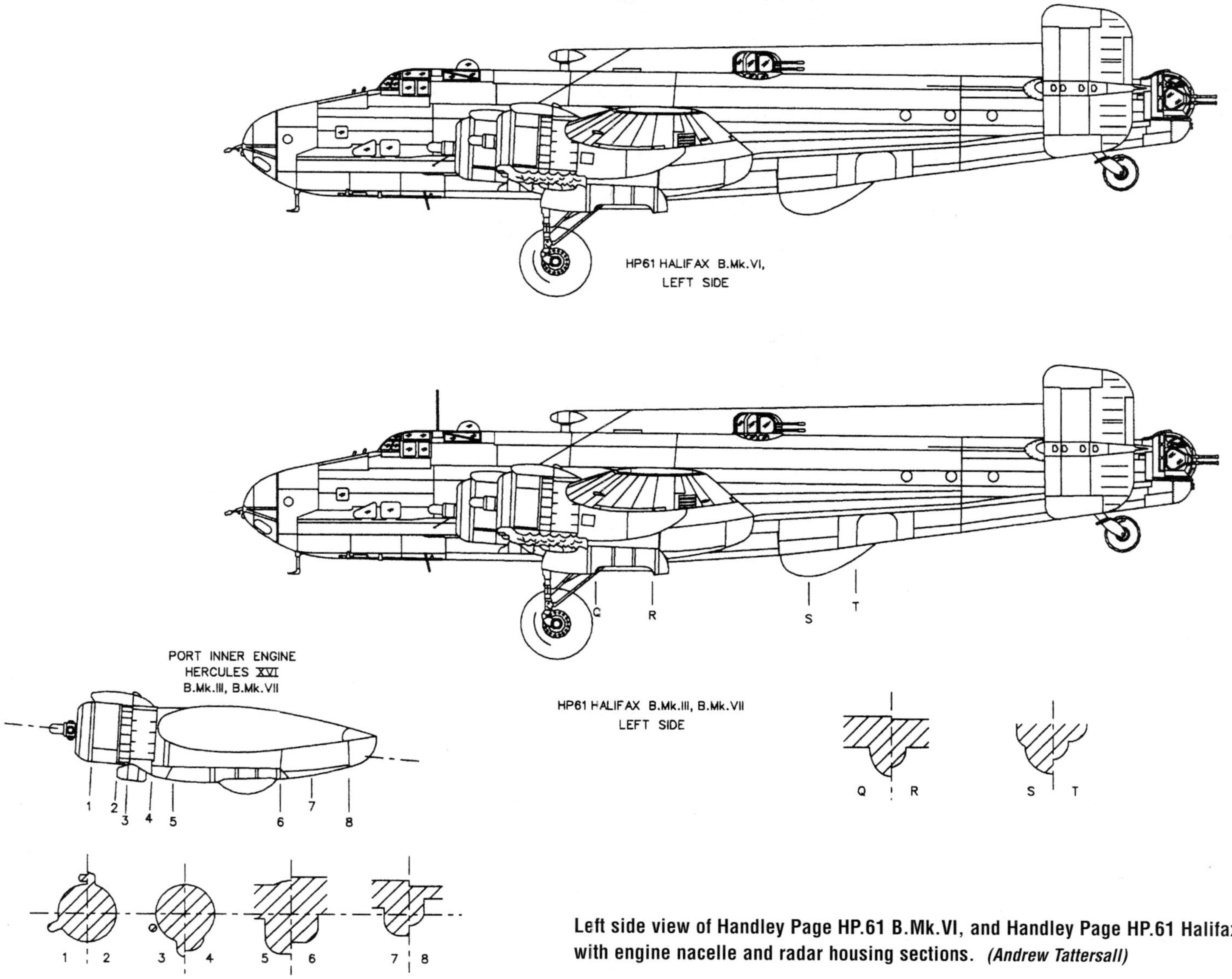

HP61 HALIFAX B.Mk.VI, LEFT SIDE

HP61 HALIFAX B.Mk.III, B.Mk.VII LEFT SIDE

PORT INNER ENGINE HERCULES XVI B.Mk.III, B.Mk.VII

Left side view of Handley Page HP.61 B.Mk.VI, and Handley Page HP.61 Halifax B.Mk.III/VII with engine nacelle and radar housing sections. *(Andrew Tattersall)*

CHAPTER FOUR
Halifax Squadrons and Units

A pair of Handley Page HP.61 Halifax B.Mk.III aircraft of the No. 6 (RCAF) Group sit in readiness for another night mission as dusk settles on their base in Yorkshire, 3 April 1945. *(CF PL-43201)*

NO. 6 (RCAF) GROUP, BOMBER COMMAND

The Handley Page Halifax served with every Royal Canadian Air Force bomber squadron based in England in the Second World War. In all, some fifteen Heavy Bomber Squadrons and three Heavy Conversion Units (HCUs) operated the various Marks of the Halifax at some time during their service.

In all, more than 232,500 men and 17,000 women served in the RCAF, both in home defence and overseas. The biggest and costliest Canadian air commitment was in Bomber Command.

The first Canadian bomber squadrons, formed in the spring of 1941, were 405(B) "Vancouver" Squadron, flying Vickers Wellington B.Mk.II aircraft (which later became the only Canadian squadron in the No. 8

Handley Page Halifax

Unit crest of No. 6 (RCAF) Group, Bomber Command

(Pathfinder) Group), and 408(B) "Goose" Squadron, flying Handley Page Hampden B.Mk.I aircraft. In December of that year, 419(B) "Moose", and 420(B) "Snowy Owl" Squadrons were organized.

W/C Johnny Fauquier was the first RCAF officer to lead a bomber squadron. He took command of 405(B) "Vancouver" Squadron in February 1942. Awarded the DFC and three times the DSO, he became the RCAF's outstanding bomber leader of the war.

During 1942, seven new RCAF bomber squadrons were formed. All were initially equipped with Vickers Wellington medium bombers. On 25 October 1942 No. 6 (RCAF) Group, Bomber Command, was formed at Linton-on-Ouse, Yorkshire. On 6 December 1942 the Group moved to permanent headquarters at Allerton Park, a seventy-five room mansion on a 200-acre estate east of Knaresborough, Yorkshire. Allerton Park Castle had been requisitioned from Lord Mowbray by the Air Ministry.

The castle was converted into offices that became the administrative and operational headquarters of No. 6 Group, which officially reached operational status on 1 January 1943. Eleven Canadian bomber squadrons were brought together to form an all-Canadian No. 6 (RCAF) Bomber Group, under the command of Air Vice-Marshal G.E. Brooks, whom Air Vice-Marshal C.M. McEwen succeeded a year later. The Canadian bomber squadrons from that date reported directly to Bomber Command Headquarters through the newly formed group. The Canadian government assumed full financial responsibility for the maintenance and administration of the group, while the aircraft, until the introduction of the Canadian built Avro Lancaster B.Mk.X, were provided by the Air Ministry.

Group Headquarters was responsible for forwarding instructions on operational requirements to the squadrons. These included mission routing, configuration of bomb load, time over target and bombing height. The stations were responsible for accommodation and feeding of personnel, and maintenance of squadron aircraft.

Initially No. 6 (RCAF) Group (composed of Squadrons 405, 408, 419, 420, 424, 425, 426, 427, 428, 429 and 431) suffered a grim casualty rate. Between 5 March and 24 June 1943 the group lost 100 aircraft, seven percent of its strength. However, by mid 1944 with better equipment and training, more experience, and fighter protection up to the targets, the situation was brought in hand. At the end of 1944, No. 6 (RCAF) Group could boast the lowest casualties of any group in Bomber Command.

Three RCAF bomber squadrons were deployed to the Mediterranean theatre. In mid 1943, 420(B), 424(B) and 425(B) Squadrons under G/C C.R. Dunlap, organized as 331 (RCAF) Wing, moved from Yorkshire to Tunisia in North Africa to support the Allied invasion of Sicily and Italy. These squadrons returned to England and rejoined No. 6 (RCAF) Group in late autumn of that year.

In 1943 three more Canadian bomber squadrons were formed, No. 432(B), No. 434(B) and No. 433(B) Squadrons.

In 1944 all RCAF squadrons had been equipped with Handley Page Halifax or Avro Lancaster aircraft. In July 1944 No. 6 Group gained 415(B) "Swordfish" Squadron for a total of fourteen heavy bomber squadrons, and 405(B) "Vancouver" Squadron returned to No. 6 (RCAF) Group in May 1945.

P/O A.C. Mynarski won a posthumous Victoria Cross during a night mission in June 1944. A mid-upper gunner in a 419(B) "Moose" Squadron Lancaster, he went to the aid of the rear gunner when his aircraft was shot down in flames. Unable to free his comrade, and with his own clothing and parachute on fire, he reluctantly bailed out.

CHAPTER FOUR

His burns were so severe that he died shortly after reaching the ground. The rear gunner survived the crash to report Mynarski's heroism.

During twenty-eight months on operations No. 6 (RCAF) Group flew 271,981 hours on 40,822 sorties and dropped 126,122 tons of bombs and mines. It lost 814 crews (six men per crew). Of the 50,000 RCAF personnel that served in Bomber Command, 10,000 gave their lives.

Out of a total of 37,000 operational missions flown in Halifax aircraft in the Second World War, 29,000 were flown by Canadian crews.

Commanding Officers of No. 6 (RCAF) Group

A/VM G.E. Brookes — 1 January 1943 to 28 February 1944
A/VM C.M. McEwen — 29 February 1944 to 13 July 1945

A group of airmen pose in front of their HP.61 Halifax B.Mk.III to promote the sale of Victory Bonds to support the Canadian war effort, 23 February 1944. *(CF)*

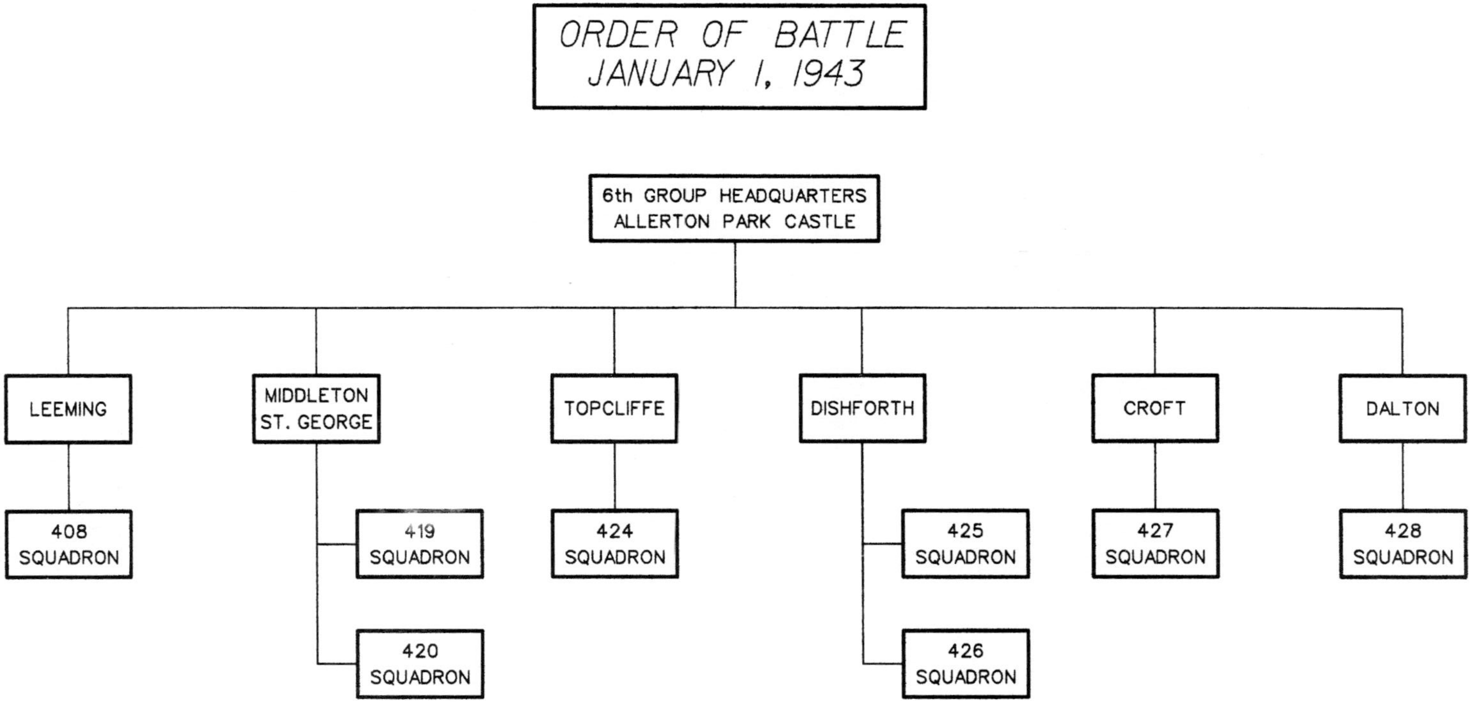

ORDER OF BATTLE JANUARY 1, 1943

- 6th GROUP HEADQUARTERS ALLERTON PARK CASTLE
  - LEEMING
    - 408 SQUADRON
  - MIDDLETON ST. GEORGE
    - 419 SQUADRON
    - 420 SQUADRON
  - TOPCLIFFE
    - 424 SQUADRON
  - DISHFORTH
    - 425 SQUADRON
    - 426 SQUADRON
  - CROFT
    - 427 SQUADRON
  - DALTON
    - 428 SQUADRON

No. 6 (RCAF) Group Order of Battle, 1 January 1943

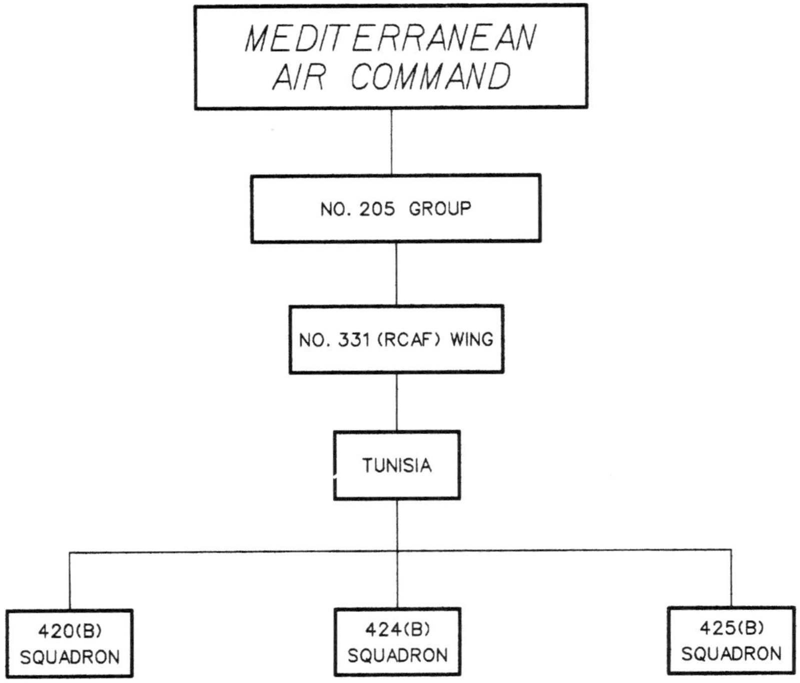

```
MEDITERRANEAN
AIR COMMAND
     |
NO. 205 GROUP
     |
NO. 331 (RCAF) WING
     |
TUNISIA
     |
 ┌───┼───┐
420(B)  424(B)  425(B)
SQUADRON SQUADRON SQUADRON
```

No. 331 (RCAF) Wing Order of Battle, 23 June 1943

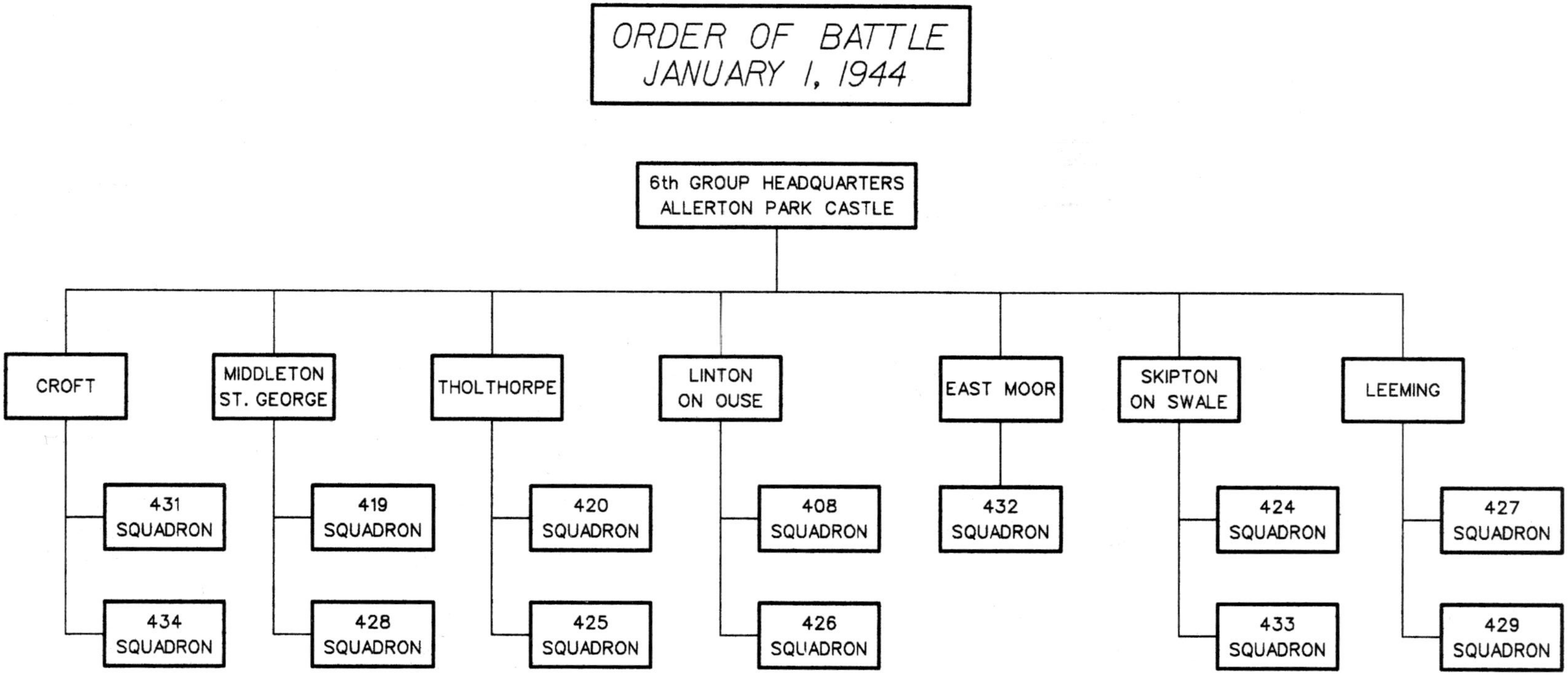

ORDER OF BATTLE JANUARY 1, 1944

- 6th GROUP HEADQUARTERS ALLERTON PARK CASTLE
  - CROFT
    - 431 SQUADRON
    - 434 SQUADRON
  - MIDDLETON ST. GEORGE
    - 419 SQUADRON
    - 428 SQUADRON
  - THOLTHORPE
    - 420 SQUADRON
    - 425 SQUADRON
  - LINTON ON OUSE
    - 408 SQUADRON
    - 426 SQUADRON
  - EAST MOOR
    - 432 SQUADRON
  - SKIPTON ON SWALE
    - 424 SQUADRON
    - 433 SQUADRON
  - LEEMING
    - 427 SQUADRON
    - 429 SQUADRON

No. 6 (RCAF) Group Order of Battle, 1 January 1944

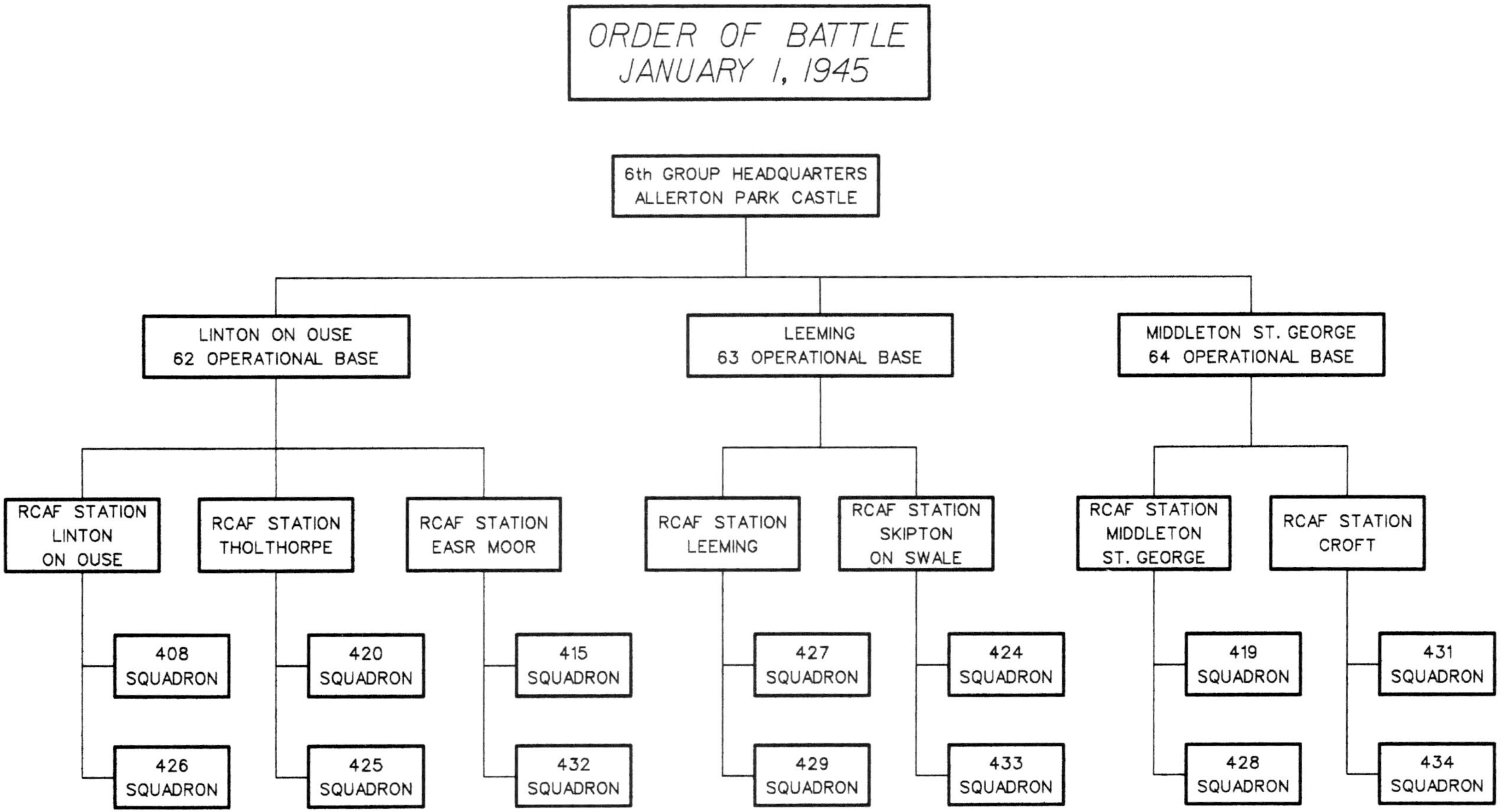

ORDER OF BATTLE JANUARY 1, 1945

- 6th GROUP HEADQUARTERS
  ALLERTON PARK CASTLE
  - LINTON ON OUSE
    62 OPERATIONAL BASE
    - RCAF STATION LINTON ON OUSE
      - 408 SQUADRON
      - 426 SQUADRON
    - RCAF STATION THOLTHORPE
      - 420 SQUADRON
      - 425 SQUADRON
    - RCAF STATION EASR MOOR
      - 415 SQUADRON
      - 432 SQUADRON
  - LEEMING
    63 OPERATIONAL BASE
    - RCAF STATION LEEMING
      - 427 SQUADRON
      - 429 SQUADRON
    - RCAF STATION SKIPTON ON SWALE
      - 424 SQUADRON
      - 433 SQUADRON
  - MIDDLETON ST. GEORGE
    64 OPERATIONAL BASE
    - RCAF STATION MIDDLETON ST. GEORGE
      - 419 SQUADRON
      - 428 SQUADRON
    - RCAF STATION CROFT
      - 431 SQUADRON
      - 434 SQUADRON

NO.6 (RCAF) Group Order of Battle, 1 January 1945

Handley Page Halifax

NO. 405 (BOMBER) "VANCOUVER" SQUADRON RCAF
UNIT CODE LQ

The first RCAF overseas bomber squadron, No. 405(B) Squadron was formed at Driffield, Yorkshire on 23 April 1941. Operating in No. 4 Group of RAF Bomber Command, the squadron was equipped with the Vickers Wellington B.Mk.II medium bomber, and carried out the first RCAF bombing mission on the night of 12/13 June. On 20 June 1941 the squadron was relocated to Pocklington, Yorkshire.

In preparation for the conversion to the Handley Page Halifax four-engine heavy bomber, a conversion flight was formed on 29 April 1942, with Halifax B.Mk.I and B.Mk.II aircraft. The unit moved to Topcliffe on 7 August 1942, was disbanded on 7 October 1942 and absorbed into 1659 Heavy Conversion Unit.

Operations with the Halifax B.Mk.II Series I commenced on 30/31 May 1942, when 405 Squadron participated in the first 1000 Bomber raid against Cologne, Germany. In August that year the unit moved to Topcliffe.

The squadron was transferred to 18 Group, RAF Coastal Command on anti-submarine and escort operations in October 1942, and operated from Beaulieu,

Unit crest of No. 405 (Bomber) "Vancouver" Squadron RCAF

Handley Page HP.59 Halifax B.Mk.II Series I in service with 405(B) Squadron in 1942. This aircraft, named the Ruhr Valley Express, had a depiction of a steam engine and tender on the nose. A new rail car would be added to the train for each operation flown. *(CF PL-10458)*

66

CHAPTER FOUR

Hampshire to assist in the protection of convoys sailing to the invasion of North Africa in Operation Torch. On its return to Bomber Command in March 1943, the squadron was assigned to No. 6 (RCAF) Group. Although initially based again at Topcliffe, in short order 405 Squadron moved to Leeming in Yorkshire.

In April 1943 405(B) Squadron was selected as the No. 6 Group representative in the RAF Bomber Command Pathfinder Force, later redesignated No. 8 (Pathfinder Force) Group. The unit was relocated at Gransden Lodge, Bedfordshire. In August 1943 the conversion to the Avro Lancaster B.Mk.I began. The last operations with the Halifax aircraft took place on 16 September, and the squadron operated the Lancaster until the war ended.

The Squadron was the first to introduce the Canadian built Lancaster B.Mk.X, when KB700 was taken on strength on 30 October 1943. The aircraft served with 405(B) Squadron until 20 December before being transferred to 419(B) "Moose" Squadron early in 1944.

On 26 May 1945 405(B) Squadron returned to No. 6 Group at Linton-on-Ouse, Yorkshire. After converting to Canadian built Avro Lancaster B.Mk.X aircraft, the squadron flew their aircraft back to Canada. Based at Greenwood, Nova Scotia, the squadron was attached to No. 664 (RCAF) Heavy Bomber Wing, Eastern Air Command and began training for the invasion of Japan as part of RAF Tiger Force. Conversion to Avro Lincoln B.Mk.XV heavy bombers was curtailed by the end of the war with Japan in August 1945. The squadron was disbanded at Greenwood on 5 September 1945.

HIGHER FORMATIONS AND BASES

BASE OF OPERATIONS	DURATION
BOMBER COMMAND	
No. 4 Group	
Driffield, Yorkshire	23 April 1941 to 19 June 1941
Pocklington, Yorkshire	20 June 1941 to 6 August 1942
Topcliffe, Yorkshire	7 August 1942 to 24 October 1942
to Coastal Command, No. 18 Group	
COASTAL COMMAND	
No. 18 Group	
Beaulieu, Hants.	25 October 1942 to 28 February 1943
to Bomber Command No. 6 (RCAF) Group	
BOMBER COMMAND	
No. 6 Group	
Topcliffe, Yorkshire	1 March 1943 to 5 March 1943
Leeming, Yorkshire	6 March 1943 to 18 April 1943
to No. 8 (Pathfinder) Group	
No. 8 (Pathfinder) Group	
Gransden Lodge, Bedfordshire	19 April 1943 to 25 May 1945
to No. 6 (RCAF) Group	

Handley Page Halifax

HIGHER FORMATIONS AND BASES (cont'd)

BASE OF OPERATIONS	DURATION
BOMBER COMMAND (cont'd) No. 6 (RCAF) Group No. 62 (RCAF) Base Linton-on-Ouse, Yorkshire *en route to Canada*	26 May 1945 to 15 June 1945 *16 June 1945 to 20 June 1945*
TIGER FORCE RCAF Eastern Air Command No. 6 (RCAF) Group No. 664 (RCAF) Heavy Bomber Wing Greenwood, Nova Scotia Disbanded at Greenwood	21 June 1945 to 5 September 1945 5 September 1945

REPRESENTATIVE AIRCRAFT TYPES

AIRCRAFT TYPE	ENGINE TYPE	DATES OF SERVICE
Vickers Wellington B.Mk.II W5553 LQ*D	Rolls Royce Merlin	May 1941 to April 1942
Handley Page H.P.57 Halifax B.Mk.I R9363 LQ*U	Rolls Royce Merlin	January 1942 to April 1942
Handley Page H.P.59 Halifax B.Mk.II W7710 LQ*R "Ruhr Valley Express"	Rolls Royce Merlin	April 1942 to September 1943
Handley Page H.P.59 Halifax B.Mk.II (Special) W1173 LQ*X	Rolls Royce Merlin	April 1942 to September 1943
Handley Page H.P.59 Halifax B/GR.Mk.II DT487 LQ*M	Rolls Royce Merlin	September 1942 to September 1943
Avro Lancaster B.Mk.I ME445 LQ*U	Rolls Royce Merlin	August 1943 to May 1945
Avro Lancaster B.Mk.III PB229 LQ*H "Honky-Tonk"	Packard Merlin	August 1943 to May 1945
Avro Lancaster B.Mk.X KB700 LQ*Q "Ruhr Express"	Packard Merlin	October 1943 to December 1943 May 1945 to September 1945

CHAPTER FOUR

NO. 408 (BOMBER) "GOOSE" SQUADRON RCAF
UNIT CODE EQ

No. 408(B) Squadron was formed on 24 June 1941 at Lindholme, Yorkshire, and operated in No. 5 Group of RAF Bomber Command. The unit was equipped with Handley Page Hampden B.Mk.I aircraft. The squadron moved to Syerston, Nottinghamshire in July. The first operation took place on 11/12 August, when four Hampden aircraft were tasked with bombing the docks at Rotterdam in the Netherlands. The squadron then moved to Balderton, Nottinghamshire in December of that year. Between January and March of 1942, the squadron was relocated to North Luffenham, Nottinghamshire while the runways at Balderton were being resurfaced.

On 16 May 1942 408(B) Conversion Flight was formed at Syerston to introduce the Avro Manchester B.Mk.IA into service, but cancelled on 19 June 1942 when the performance of the aircraft proved unacceptable. The flight was reformed on 20 September 1942 at Leeming, Yorkshire with Handley Page Halifax B.Mk.I, B.Mk.II, and B.Mk.V aircraft. The unit was disbanded on 7 October 1942 and absorbed into 1659 Heavy Conversion Unit.

Unit Crest of No. 408 (Bomber) "Goose" Squadron RCAF

Handley Page HP.61 Halifax B.Mk.III aircraft of 408(B) "Goose" Sqn RCAF at dispersal at Linton-on-Ouse, Yorkshire, autumn 1944. *(CF PL-43480)*

Handley Page Halifax

The squadron moved once again, in September 1942, this time to Leeming to operate under No. 4 Group of RAF Bomber Command. Although the unit was initially equipped with Halifax B.Mk.V aircraft, Halifax B.Mk.II Series I were taken on strength in December 1942 It was at Leeming, on 1 January 1943, that the squadron transferred to the new No. 6 (RCAF) Group.

When in August 1943 the squadron was relocated to Linton-on-Ouse, Yorkshire, they traded their Halifax aircraft for the Avro Lancaster B.Mk.II. In September 1944 408 was again equipped with the Handley Page Halifax, this time with the Hercules powered B.Mk.III and B.Mk.VII aircraft. Its last operational mission took place on 25 April 1945.

At the end of hostilities in Europe in May 1945 the Canadian built Avro Lancaster B.Mk.X aircraft were taken on strength. The squadron flew the Lancaster aircraft back to Canada. Based at Greenwood, Nova Scotia, the squadron was attached to No. 664 (RCAF) Heavy Bomber Wing, and began training for participation in the invasion of Japan as part of Tiger Force. The anticipated conversion to Avro Lincoln B.Mk.XV bombers was curtailed by the end of the war with Japan in August 1945. The squadron disbanded at Greenwood on 5 September 1945.

HIGHER FORMATIONS AND BASES
(For example Bomber Command was a higher formation to No. 5 Group)

BASE OF OPERATIONS	DURATION
BOMBER COMMAND	
No. 5 Group	
Lindholme, Yorkshire	24 June 1941 to 19 July 1941
Syerston, Nottinghamshire	20 July 1941 to 8 December 1941
Balderton, Nottinghamshire	9 December 1941 to 12 September 1942
North Luffenham, Nottinghamshire	*25 January 1942 to 17 March 1942*
(Balderton runways under repair)	
to No. 4 Group	
No. 4 Group	
Leeming, Yorkshire	14 September 1942 to 31 December 1942
to No. 6 (RCAF) Group	
No. 6 (RCAF) Group	
Leeming, Yorkshire	1 January 1943 to 26 August 1943
No. 62 (RCAF) Base	
Linton-on-Ouse, Yorkshire	27 August 1943 to 13 June 1945
en route to Canada	*14 June 1945 to 17 June 1945*
TIGER FORCE	
RCAF Eastern Air Command	
No. 6 (RCAF) Group	
No. 664 (RCAF) Heavy Bomber Wing	
Greenwood, Nova Scotia	18 June 1945 to 5 September 1945
Disbanded at Greenwood	5 September 1945

CHAPTER FOUR

REPRESENTATIVE AIRCRAFT TYPES

AIRCRAFT TYPE	ENGINE TYPE	DATES OF SERVICE
Handley Page Hampden B.Mk.I AE269 EQ*W	Bristol Pegasus	July 1941 to September 1942
Avro Manchester B.Mk.Ia L7401 EQ*X	Rolls Royce Vulture	May 1942 to June 1942
Handley Page H.P.57 Halifax B.Mk.I L9524 EQ*S	Rolls Royce Merlin	September 1942 to December 1942
Handley Page H.P.63 Halifax B.Mk.V DG231	Rolls Royce Merlin	September 1942 to December 1942
Handley Page H.P.59 Halifax B/GR Mk.II JB969 EQ*D	Rolls Royce Merlin	December 1942 to October 1943
Avro Lancaster B.Mk.II DS858 EQ*D	Bristol Hercules	October 1943 to September 1944
Handley Page H.P.61 Halifax B.Mk.III NR124 EQ*S	Bristol Hercules	September 1944 to May 1945
Handley Page H.P.61 Halifax B.Mk.VII PN230 EQ*V "Vicky The Vicious Virgin"	Bristol Hercules	September 1944 to May 1945
Avro Lancaster B.Mk.X KB996 EQ*P	Packard Merlin	May 1945 to September 1945

NO. 415 (BOMBER) "SWORDFISH" SQUADRON RCAF
UNIT CODE   (COASTAL) GX / NH
(BOMBER) 6U

Formed in England as No. 415(TB) Squadron on 20 August 1941 at Thorney Island, Hampshire, it operated in No. 16 Group of RAF Coastal Command. The squadron was initially equipped with Bristol Beaufort torpedo bombers and Bristol Blenheim Mk.IVF coastal fighters for operational training.

After conversion to Handley Page Hampden TB.Mk.I torpedo bombers in April 1942 the first anti-shipping operation was flown on the 27th of that month. The squadron had relocated to St. Eval, Cornwall where it operated in No. 19 Group until 15 May when it once again took up station at Thorney Island in 16 Group. On 29 May three Hampdens attacked a convoy, damaging a ship, but losing one aircraft to flak. The squadron moved to North Coates, Lincolnshire, in June.

Unit crest of No. 415 (Bomber) "Swordfish" Squadron RCAF

Handley Page Halifax

A Handley Page HP.61 Halifax B.Mk.VII of 415(B) "Swordfish" Sqn RCAF taxies for takeoff, East Moor, Yorkshire, 9 April 1945. *(CF PL-43479)*

Following a move to Scotland, the squadron operated in No. 18 Group, from Wick, Caithness, during July and August; Tain, Rosshire, briefly in early September; and Leuchars, Fife, from September to November. One detachment of six aircraft operated in October from St. Eval in No. 19 Group, and another six from Thorney Island, in No. 16 Group during October and November.

During the time the unit operated the Handley Page Hampden TB.Mk.I aircraft, anti-shipping operations were conducted in the English Channel and along the coastal waters of France, Belgium and the Netherlands. As operations with the Hampden aircraft became increasingly hazardous, the squadron changed to night operations.

Thorney Island became the squadron's base for a year until November 1943. In September, the unit had been re-equipped with a mixed complement of Fairey Albacore TB.Mk.I aircraft and Vickers Wellington GR.Mk.XIII. They continued with nighttime anti-shipping operations, largely against the German Navy E-Boats. The Wellington aircraft located and illuminated the targets with their Leigh lights, while the vessels were attacked by the Albacores. Operations in conjunction with Beaufighter units were also conducted. From November 1943 until July 1944 the unit operated from Bircham Newton, Norfolk, still in No. 16 Group. From May to July 1944 detachments of Albacores also operated from Manston, Kent, and Thorney Island in Hampshire, and Winkleigh in Devonshire. Also, Wellington detachments operated from Docking, Norfolk, and North Coates, Lincolnshire. After D-Day, as enemy shipping operations in the English Channel were drastically curtailed, the squadron handed over its Albacore aircraft to No. 119 Squadron, RAF.

Redesignated No. 415(B) on 12 July 1944, the squadron was reformed as the fifteenth and final Canadian bomber squadron, using trained personnel from within No. 6 (RCAF) Group. Relocated to East Moor, Yorkshire, the squadron was equipped with Handley Page Halifax B.Mk.III and later, some Handley Page B.Mk.VII aircraft. The first operational mission was carried out on 28 July 1944 when aircraft were despatched to bomb Hamburg, Germany. The last operational mission took place on 25 April 1945 when eighteen aircraft bombed gun positions on Wangerooge Island. The squadron was disbanded at East Moor on 15 May 1945.

CHAPTER FOUR

HIGHER FORMATIONS AND BASES

BASE OF OPERATIONS	DURATION
COASTAL COMMAND	
No. 16 Group Thorney Island, Hampshire	26 August 1941 to 10 April 1942
No. 19 Group St. Eval, Cornwall	11 April 1942 to 15 May 1942
No. 16 Group Thorney Island, Hampshire North Coates, Lincolnshire	16 May 1942 to 4 June 1942 5 June 1942 to 30 July 1942
No. 18 Group Wick, Caithness Tain, Rosshire Leuchars, Fife	31 July 1942 to 31 August 1942 1 September 1942 to 9 September 1942 10 September 1942 to 10 November 1942
No. 19 Group St. Eval, Cornwall (6 aircraft)	18 October 1942 to 31 October 1942
No. 16 Group Thorney Island, Hampshire (6 aircraft)	18 October 1942 to 10 November 1943
No. 16 Group Thorney Island, Hampshire Bircham Newton, Norfolk	11 November 1942 to 14 November 1942 15 November 1943 to 11 July 1944
BOMBER COMMAND	
No. 6 (RCAF) Group Bircham Newton, Norfolk	12 July 1944 to 25 July 1944
No. 62 (RCAF) Base East Moor, Yorkshire Disbanded at East Moor	26 July 1944 to 15 May 1945 15 May 1945

Handley Page Halifax

REPRESENTATIVE AIRCRAFT TYPES

AIRCRAFT TYPE	ENGINE TYPE	DATES OF SERVICE
COASTAL COMMAND		
Bristol Beaufort TB.Mk.I L9802 GX*S	Bristol Taurus	September 1941 to February 1942
Bristol Blenheim Mk.IVF L9476 GX*Y	Bristol Mercury	December 1941 to February 1942
Handley Page Hampden TB.Mk.I AT243 GX*N	Bristol Pegasus	January 1942 to September 1943
Handley Page Hampden TB.Mk.II X3115 GX*J	Wright Cyclone	January 1942 to September 1943
Vickers Wellington GR.Mk.XII MF639 NH*H	Bristol Hercules	September 1943 to July 1944
Vickers Wellington GR.Mk.XIII NZ756 NH*G	Bristol Hercules	September 1943 to July 1944
Fairey Albacore TB.Mk.I X9140 NH*X1	Bristol Taurus	October 1943 to July 1944
BOMBER COMMAND		
Handley Page Halifax B.Mk.III MZ861 6U*Z	Bristol Hercules	July 1944 to May 1945
Handley Page Halifax B/A.Mk.III NA608 6U*H	Bristol Hercules	July 1944 to May 1945
Handley Page Halifax B/A Met. Mk.III NA186 6U*U	Bristol Hercules	March 1944 to May 1945
Handley Page Halifax B.Mk.VII RG447 6U*S	Bristol Hercules	March 1944 to May 1945

NO. 419 (BOMBER) "MOOSE" SQUADRON RCAF
UNIT CODE VR

Formed in England as No. 419(B) Squadron on 15 December 1941 at Mildenhall, Suffolk, and operating in No. 3 Group of RAF Bomber Command, the unit was equipped with Vickers Wellington B.Mk.IC and B.Mk.III aircraft. The first operational sortie was made on 11 January 1942, when two aircraft bombed Brest, France.

The squadron relocated to Leeming, Yorkshire, as part of No. 4 Group in August 1942. Later in August it moved again, to Topcliffe, and to Croft in October 1942. In October, as well, the squadron was re-equipped with Handley Page Halifax B.Mk.II heavy bomber aircraft.

Another move was made to Middleton St. George, Durham in November 1942, where the squadron was based for the remainder of the war. On 1 January 1943 the squadron was transferred to the newly formed No. 6 (RCAF) Group. In April 1944 the unit was re-equipped with Canadian built Avro Lancaster B.Mk.X heavy bombers.

On 12 June 1944, during a raid on railway marshalling yards at Cambrai, France, a Lancaster B.Mk.X bomber of 419(B) Squadron, serial number KB726, coded VR*A, was attacked by a Luftwaffe Junkers 88C-6 night fighter. The captain ordered the crew to

CHAPTER FOUR

abandon the aircraft which had been set aflame in the attack. The rear gunner was trapped in his turret, and the mid-upper gunner, P/O Andrew Mynarski, made every effort to free him. Finally, with his own clothing and parachute aflame, he bailed out of the burning bomber. Unfortunately, he died in a German hospital of his burns and injuries sustained in the low-level bailout. The tail gunner survived the crash and was taken prisoner of war. After the war, when details of his attempt to free the trapped tail gunner became known, P/O Andrew Mynarski was posthumously awarded the Victoria Cross.

After the end of hostilities in Europe, the squadron flew back to Canada. Based at Yarmouth, Nova Scotia, the unit was attached to 661 (RCAF) Heavy Bomber Wing, training for operations with Tiger Force and the war against Japan in the Far East. The anticipated re-equipment with Avro Lincoln B.Mk.XV heavy bombers was cancelled when the war with Japan was ended in August 1945. The squadron was disbanded at Yarmouth on 5 September 1945.

Unit crest of No. 419 (Bomber) "Moose" Squadron RCAF

HIGHER FORMATIONS AND BASES

BASE OF OPERATIONS	DURATION
BOMBER COMMAND	
No. 4 Group	
Leeming, Yorkshire	13 August 1942 to 17 August 1942
Topcliffe, Yorkshire	18 August 1942 to 30 September 1942
Croft, Yorkshire	1 October 1942 to 9 November 1942
Middleton St. George, Durham	10 November 1942 to 31 December 1942
to No. 6 (RCAF) Group	
No. 6 (RCAF) Group	
No. 64 (RCAF) Base	
Middleton St. George, Durham	1 January 1943 to 31 May 1943
en route to Canada	*1 June 1945 to 4 June 1945*
TIGER FORCE	
RCAF Eastern Air Command	
No. 6 (RCAF) Group	
No. 661 (RCAF) Heavy Bomber Wing	
Yarmouth, Nova Scotia	5 June 1945 to 5 September 1945
Disbanded at Yarmouth	5 September 1945

Handley Page Halifax

REPRESENTATIVE AIRCRAFT TYPES

AIRCRAFT TYPE	ENGINE TYPE	DATES OF SERVICE
Vickers Wellington B.Mk.IC	Bristol Pegasus	January 1942 to November 1942
Vickers Wellington B.Mk.III	Bristol Hercules	February 1942 to November 1942
Handley Page H.P.59 Halifax B/GR.Mk.II DT669 VR*L "London II"	Rolls Royce Merlin	November 1942 to April 1944
Handley Page H.P.59 Halifax B.Mk.II W1271 VR*P	Rolls Royce Merlin	January 1943 to April 1944
Avro Lancaster B.Mk.X KB726 VR*A	Packard Merlin	April 1944 to September 1945

Posed in front of their Handley Page HP.59 Halifax B.Mk.II Series I (Special) named "The Mermaid" are members of 419(B) Sqn with a bomb decorated to mark the occasion of the squadron's 1,000th sortie, 6 July 1943.
(CF PL-19060)

CHAPTER FOUR

NO. 420 (BOMBER) "SNOWY OWL" SQUADRON RCAF
UNIT CODE PT

No. 420(B) Squadron was formed on 19 December 1941 at Waddington, Lincolnshire, and operated in No. 5 Group of RAF Bomber Command. The unit was equipped with Handley Page Hampden B.Mk.I aircraft. The squadron flew its first operational sortie on the night of 21 January 1942.

A conversion flight was formed in April 1942 to introduce the Avro Manchester B.Mk.IA into service, but cancelled in June when the performance of the aircraft proved unacceptable, as had been experienced by 408(B) Squadron. When it moved to Skipton-on-Swale and No. 4 Group in August 1942 the squadron was re-equipped with Vickers Wellington B.Mk.III, and later B.Mk.X aircraft.

Another move was made in October 1942 to Middleton St. George, where in January 1943 it joined the newly formed No. 6 (RCAF) Group. However, in May 1943 the squadron was posted to North Africa as part of 331 (RCAF) Wing, No. 205 Group in Mediterranean Air Command. From its bases in Tunisia, the squadron flew tropicalized Vickers Wellington B.Mk.III aircraft from June to October in support of the invasions of Sicily and Italy.

In November, the squadron returned to No. 6 (RCAF) Group. Based at Dalton, Yorkshire, the squadron again operated Vickers Wellington B.Mk.X aircraft. When the unit relocated to Tholthorpe, Yorkshire, in December a conversion was made to Handley Page Halifax B.Mk.III, and later B.Mk.VII aircraft, which were flown until the hostilities in Europe ended in May 1945.

Unit crest of No. 420 (Bomber) "Snowy Owl" Squadron RCAF

Conversion to Canadian built Avro Lancaster B.Mk.X aircraft was started in April 1945. In June, the squadron flew their aircraft back to Canada. Based at Debert, Nova Scotia, the squadron began training for operations with Tiger Force in support of the invasion of Japan. The anticipated re-equipment with Avro Lincoln B.Mk.XV aircraft was cut short when the war with Japan ended in August 1945. The squadron disbanded at Debert on 5 September 1945.

A Handley Page HP.61 Halifax B.Mk.III of 420(B) "Snowy Owl" Sqn RCAF positions for takeoff, with a mobile control unit and other vehicles in the foreground. *(CF PL-41612)*

Handley Page Halifax

HIGHER FORMATIONS AND BASES

BASE OF OPERATIONS	DURATION
BOMBER COMMAND	
No. 5 Group	
No. 420 (Bomber) Squadron RCAF	
Waddington, Lincolnshire	19 December 1942 to 6 August 1942
to No. 4 Group	
No. 4 Group	
Skipton-on-Swale, Yorkshire	7 August 1942 to 15 October 1942
Middleton St. George, Durham	10 November 1942 to 31 December 1942
to No. 6 (RCAF) Group	
No. 6 (RCAF) Group	
Middleton St. George, Durham	1 January 1943 to 15 May 1943
en route to North Africa, No. 205 Group	*16 May 1943 to 18 June 1943*
MEDITERRANEAN AIR COMMAND	
No. 205 Group	
No. 331 (RCAF) Wing	
Kairouan/Zina, Tunisia	19 June 1943 to 29 September 1943
Hani East Landing Ground, Tunisia	29 September 1943 to 16 October 1943
en route to England (minus aircraft)	*17 October 1943 to 5 November 1943*
BOMBER COMMAND	
No. 6 (RCAF) Group	
No. 61 (RCAF) Base	
Dalton, Yorkshire	6 November 1943 to 11 December 1943
No. 62 (RCAF) Base	
Tholthorpe, Yorkshire	12 December 1943 to 11 June 1945
en route to Canada	*12 June 1945 to 16 June 1945*
TIGER FORCE	
RCAF Eastern Air Command	
No. 6 (RCAF) Group	
No. 663 (RCAF) Heavy Bomber Wing	
Debert, Nova Scotia	16 June 1945 to 5 September 1945
Disbanded at Debert	5 September 1945

CHAPTER FOUR

REPRESENTATIVE AIRCRAFT TYPES

AIRCRAFT TYPE	ENGINE TYPE	DATES OF SERVICE
Handley Page Hampden B.Mk.I AD915 PT*F	Bristol Pegasus	December 1941 to August 1942
Avro Manchester B.Mk.IA L7402	Rolls Royce Vulture	April 1942 to June 1942
Vickers Wellington B.Mk.III BK365 PT*L	Bristol Hercules	August 1942 to April 1943
Vickers Wellington B.Mk.X HE975 PT*U	Bristol Hercules	February 1943 to May 1943 November 1943 to December 1943
Vickers Wellington B.Mk.III (Trop.)	Bristol Hercules	May 1943 to October 1943
Handley Page H.P.61 Halifax B.Mk.III MZ423 PT*B	Bristol Hercules	December 1943 to May 1945
Handley Page H.P.61 Halifax B.Met.Mk.III RG347	Bristol Hercules	
Handley Page H.P.61 Halifax B/A.Mk.III LW392 PT*S	Bristol Hercules	February 1944 to January 1945
Handley Page H.P. Halifax B/A Met.Mk.III NA169 PT*O	Bristol Hercules	January 1945 to March 1945
Handley Page H.P.61 HalifaxB/GR.Mk.III HX346	Bristol Hercules	November 1944
Handley Page H.P.61 Halifax B.Mk.VII LW207	Bristol Hercules	July 1944 to December 1944
Avro Lancaster B.Mk.X KB896 PT*O	Packard Merlin	April 1945 to September 1945

Handley Page Halifax

NO. 424 (BOMBER) "TIGER" SQUADRON RCAF
UNIT CODE   QB

Formed at Topcliffe, Yorkshire as No. 424(B) Squadron on 15 October 1942, the squadron operated in No. 4 Group RAF Bomber Command. It was equipped with Vickers Wellington B.Mk.III aircraft. The squadron was transferred to No. 6 (RCAF) Group in January 1943 and re-equipped with Vickers Wellington B.Mk.X aircraft They moved to Leeming in April and to Dalton, Yorkshire in May 1943.

In May 1943 the squadron was posted to North Africa as part of 331 (RCAF) Wing, No. 205 Group in Mediterranean Air Command. From bases in Tunisia, they flew tropicalized Vickers Wellington B.Mk.X aircraft from June to October in support of the invasions of Sicily and Italy.

In November 1943 the squadron returned to No. 6 (RCAF) Group. Based at Skipton-on-Swale, Yorkshire, the squadron again operated Vickers Wellington B.Mk.X aircraft. In December, they converted to Handley Page Halifax B.Mk.III aircraft, which they flew until January 1945. At that time the unit converted to Avro Lancaster B.Mk.I and B.Mk.III aircraft.

Operations continued in No. 6 Group until August, when the squadron transferred to No. 1 Group, RAF Bomber Command, and began ferrying troops back to England from Italy. The unit was disbanded at Skipton-on-Swale on 15 October 1945.

Unit Crest of No. 424 (Bomber) "Tiger" Squadron RCAF

Handley Page HP.61 Halifax B.Mk.III serial number LW119, coded QB*O of 424(B) "Tiger" Sqn RCAF taxiing into position for takeoff at Skipton-on-Swale, Yorkshire, 13 November 1944. *(CF PL-41055)*

CHAPTER FOUR

HIGHER FORMATIONS AND BASES

BASE OF OPERATIONS	DURATION
BOMBER COMMAND	
No. 4 Group	
Topcliffe, Yorkshire	15 October 1942 to 31 December 1942
to No. 6 (RCAF) Group	
No. 6 (RCAF) Group	
Topcliffe, Yorkshire	1 January 1943 to 7 April 1943
Leeming, Yorkshire	8 April 1943 to 2 May 1943
Dalton, Yorkshire	3 May 1943 to 15 May 1943
en route to North Africa, No. 205 Group	*16 May 1943 to 22 June 1943*
MEDITERRANEAN AIR COMMAND	
No. 205 Group	
No. 331 (RCAF) Wing	
Kairouan/Pavillier, Tunisia	23 June 1943 to 29 September 1943
Hani East Landing Ground, Tunisia	30 September 1943 to 18 October 1943
en route to England (minus aircraft)	26 October 1943 to 5 November 1943
BOMBER COMMAND	
No. 6 (RCAF) Group	
No. 63 (RCAF) Base	
Skipton-on-Swale, Yorkshire	6 November 1943 to 29 August 1945
to No. 1 Group	
No. 1 Group	
No. 63 (RCAF) Base	
Skipton-on-Swale, Yorkshire	30 August 1945 to 15 October 1945
Disbanded at Skipton-on-Swale	15 October 1945

Handley Page Halifax

REPRESENTATIVE AIRCRAFT TYPES

AIRCRAFT TYPE	ENGINE TYPE	DATES OF SERVICE
Vickers Wellington B.Mk.III Z1674 QB*G	Bristol Hercules	October 1942 to April 1943
Vickers Wellington B.Mk.X HE159 QB*P	Bristol Hercules	February 1943 to May 1943
Vickers Wellington B.Mk.X (Trop.) HE273 QB*B	Bristol Hercules	May 1943 to November 1943
Handley Page H.P.61 Halifax B.Mk.III LW119 QB*O "Oscar" (Nov 1944) MZ896 QB*O "Oscar"	Bristol Hercules	December 1943 to January 1945
Handley Page H.P.61 Halifax B/GR.Mk.III HX311 QB*A	Bristol Hercules	February 1944 to January 1945
Avro Lancaster B.MK.I NG347 QB*P "Piccadilly Princess"	Rolls Royce Merlin	January 1945 to October 1945
Avro Lancaster B.Mk.III RF128 QB*V	Packard Merlin	January 1945 to October 1945

NO. 425 (BOMBER) "ALOUETTE" SQUADRON RCAF
UNIT CODE   KW

Formed at Dishforth, Yorkshire as No. 425(B) Sqn on 25 June 1942 the unit operated in No. 4 Group. It was equipped with Vickers Wellington B.Mk.III aircraft. The first operational mission was flown on 5 October when five aircraft bombed Aachen, Germany.

On 1 January 1943 the squadron was transferred to the newly formed No. 6 (RCAF) Group, and in April was re-equipped with Vickers Wellington B.Mk.X aircraft. In May the squadron was posted to North Africa as part of 331 (RCAF) Wing, No. 205 Group in Mediterranean Air Command. From its bases in Tunisia, the squadron flew tropicalized Vickers Wellington B.Mk.X aircraft from June to October in support of the invasions of Sicily and Italy.

In November 1943 the squadron returned to No. 6 (RCAF) Group. Based at Dishforth, Yorkshire, the squadron again operated Vickers Wellington B.Mk.X aircraft. In December, they converted to Handley Page Halifax B.Mk.III aircraft, which they flew until May 1945. On 25 May 1945 in the last operational mission eighteen Halifax aircraft bombed gun positions on Wangerooge Island.

Unit crest of No. 425 (Bomber) "Alouette" Squadron RCAF

CHAPTER FOUR

Conversion to Canadian built Avro Lancaster B.Mk.X aircraft was started in April 1945 and in June, the squadron flew these aircraft back to Canada. Based at Debert, Nova Scotia, the squadron began training for operations with Tiger Force in support of the invasion of Japan. The anticipated re-equipment with Avro Lincoln B.Mk.XV aircraft was cut short when the war with Japan ended in August 1945. The squadron disbanded at Debert on 5 September 1945.

A Handley Page HP.61 Halifax B.Mk.III heavy bomber of 425(B) "Alouette" Sqn RCAF is marshalled into position for takeoff from its base at Tholthorpe, Yorkshire in November 1944. Note that the aircraft is equipped with the underbelly .50 calibre gun turret. *(CF PL-40557)*

Handley Page Halifax

HIGHER FORMATIONS AND BASES

BASE OF OPERATIONS	DURATION
BOMBER COMMAND	
No. 4 Group	
Dishforth, Yorkshire	25 June 1942 to 31 December 1942.
No. 6 (RCAF) Group	
Dishforth, Yorkshire	1 January 1943 to 15 May 1943
en route to North Africa, No. 205 Group	*16 May 1943 to 22 June 1943*
MEDITERRANEAN AIR COMMAND	
No. 205 Group	
No. 331 (RCAF) Wing	
Kairouan/Zina, Tunisia	23 June 1943 to 29 September 1943
Hani East Landing Ground, Tunisia	30 September 1943 to 17 October 1943
en route to England (minus aircraft)	*26 October 1943 to 5 November 1943*
BOMBER COMMAND	
No. 6 (RCAF) Group	
No. 61 (RCAF) Base	
Dishforth, Yorkshire	6 November 1943 to 9 December 1943
No. 62 (RCAF) Base	
Tholthorpe, Yorkshire	10 December 1943 to 12 June 1945
en route to Canada	*13 June 1945 to 14 June 1945*
TIGER FORCE	
RCAF Eastern Air Command	
No. 6 (RCAF) Group	
No. 663 (RCAF) Heavy Bomber Wing	
Debert, Nova Scotia	15 June 1945 to 5 September 1945
Disbanded at Debert	5 September 1945

CHAPTER FOUR

REPRESENTATIVE AIRCRAFT TYPES

AIRCRAFT TYPE	ENGINE TYPE	DATES OF SERVICE
Vickers Wellington B.Mk.III X3803 KW*H	Bristol Hercules	August 1942 to April 1943
Vickers Wellington B.Mk.X HE268 KW*K	Bristol Hercules	April 1943 to May 1943
Vickers Wellington B.Mk.X (Trop.)	Bristol Hercules	June 1943 to October 1943
Handley Page H.P.61 Halifax B.Mk.III MZ454 KW*S MZ620 KW*T (Nov 1944) with under turret	Bristol Hercules	December 1943 to May 1945
Handley Page H.P.61 Halifax B/A Mk.III LL594 KW*X	Bristol Hercules	February 1944 to November 1944
Handley Page H.P.61 Halifax B/A Met.Mk.III NA201 KW*S	Bristol Hercules	c.mid 1944
Avro Lancaster B.Mk.X KB936 KW*G	Packard Merlin	May 1945 to September 1945

NO. 426 (BOMBER) "THUNDERBIRD" SQUADRON RCAF
UNIT CODE   (BOMBER) OW
(TRANSPORT) OLW

Formed at Dishforth, Yorkshire, on 15 October 1942, No. 426(B) "Thunderbird" Squadron RCAF operated in No. 4 Group. The unit was equipped with Vickers Wellington B.Mk.III medium bombers. Its first operational mission was flown on 14 January 1943 when seven aircraft bombed Lorient, France.

On 1 January 1943 the squadron was transferred to the newly formed No. 6 (RCAF) Group, and in February began re-equipping with Vickers Wellington B.Mk.X aircraft. On 18 June 1943 the squadron moved to Linton-on-Ouse, and in July began operations on the Avro Lancaster B.Mk.II heavy bomber.

The squadron converted to the Handley Page HP.61 Halifax B.Mk.III in April 1944, and in June Halifax B.Mk.VII were added to the operational inventory. The Halifax aircraft were operated until the end of hostilities in Europe. The last Halifax operational mission was flown on 24 April 1945 when twenty aircraft bombed gun positions on the island of Wangerooge.

Unit crest of No. 426 (Bomber) "Thunderbird" Squadron RCAF

Handley Page Halifax

Handley Page HP.61 Halifax B.Mk.VII of 426(B) "Thunderbird" Sqn RCAF in its dispersal at Linton-on-Ouse, Yorkshire, 9 April 1945. *(CF PL-43473)*

To meet a requirement for long range transport units in support of the Bomber Command Tiger Force in the Pacific theatre, on 25 May 1945 the unit was redesignated No. 426(T) Squadron. Transferred to 47 Group Transport Command, and operating from Driffield, Yorkshire, the squadron was converted to Consolidated Liberator C.Mk.VI and C.Mk.VIII heavy transports.

The squadron moved to Tempsford, Bedfordshire on 25 June 1945 and was engaged in the airlift of British troops to Egypt and Indian troops from Egypt to India. Later, British troops were returned to England from Egypt.

The squadron was disbanded at Tempsford, Bedfordshire on 1 January 1946. Its Liberator aircraft were reassigned to RAF Liberator squadrons, as were personnel who volunteered for further service.

HIGHER FORMATIONS AND BASES

BASE OF OPERATIONS	DURATION
BOMBER COMMAND	
No. 4 Group	
Dishforth, Yorkshire	15 October 1942 to 31 December 1942
No. 6 (RCAF) Group	
No. 61 (RCAF) Base	
Dishforth, Yorkshire	1 January 1943 to 17 June 1943
No. 62 (RCAF) Base	
Linton-on-Ouse, Yorkshire	18 June 1943 to 24 May 1945
to Transport Command, No. 47 Group	
TRANSPORT COMMAND	
No. 47 Group	
Driffield, Yorkshire	25 May 1945 to 24 June 1945
Tempsford, Bedfordshire	25 June 1945 to 1 January 1946

CHAPTER FOUR

REPRESENTATIVE AIRCRAFT TYPES

AIRCRAFT TYPE	ENGINE TYPE	DATES OF SERVICE
BOMBER COMMAND		
Vickers Wellington B.Mk.III Z1604 OW*P	Bristol Hercules	November 1942 to March 1943
Vickers Wellington B.Mk.X	Bristol Hercules	February 1943 to May 1943
Avro Lancaster B.Mk.II DS647 OW*R	Bristol Hercules	July 1943 to May 1944
Handley Page H.P.61 Halifax B.Mk.III LK796 OW*M	Bristol Hercules	April 1944 to June 1944
Handley Page H.P.61 Halifax B/A Mk.III LW377 OW*G	Bristol Hercules	March to September 1944
Handley Page H.P.61 Halifax B.Mk.VII NP808 OW*E "I'm Easy"	Bristol Hercules	June 1944 to May 1945
Handley Page H.P.61 Halifax B.Mk.III NA202 OW*A	Bristol Hercules	December 1944 to March 1945
TRANSPORT COMMAND		
Consolidate Liberator C.Mk.VI KG918 OLW*X	Pratt & Whitney R-1830	July 1945 to December 1945
Consolidated Liberator C.Mk.VIII KL625 OLW*V	Pratt & Whitney R-1830	July 1945 to December 1945

Handley Page Halifax

NO. 427 (BOMBER) "LION" SQUADRON RCAF
UNIT CODE ZL

Formed at Croft, Yorkshire, on 7 November 1942, 427(B) "Lion" Squadron RCAF, operated in No. 3 Group. The squadron was equipped with Vickers Wellington B.Mk.III medium bombers. Their first operational mission was flown on 14 December 1942 when three aircraft were despatched on a mine laying mission in the area of the Frisian Islands.

On 1 January 1943 the squadron was transferred to the newly formed No. 6 (RCAF) Group. On 5 May 1943 a move was made to Leeming, Yorkshire and the squadron converted to Handley Page HP.63 Halifax B.Mk.V (Special) heavy bombers. In January 1944 the squadron re-equipped with Handley Page HP.61 Halifax B.Mk.IIII aircraft. The last operational mission in Halifax aircraft took place on 3 March 1945, laying mines off the coast of Norway.

In February 1945 the squadron began converting to Avro Lancaster B.Mk.I/III aircraft. It continued operations in No. 6 (RCAF) Group until the end of hostilities in Europe, and remained at Leeming where, on 30 August 1945, it transferred to No. 1 Group RAF Bomber Command Strike Force. The squadron airlifted Allied prisoners of war and British troops from Italy back to England. It was disbanded at Leeming on 1 June 1946.

Unit crest of No. 427 (Bomber) "Lion" Squadron RCAF

Handley Page HP.63 Halifax B.Mk.VA of 427(B) "Lion" Sqn RCAF being bombed up and serviced at Leeming, Yorkshire, 1 September 1943. *(CF PL-19591)*

CHAPTER FOUR

HIGHER FORMATIONS AND BASES

BASE OF OPERATIONS	DURATION
BOMBER COMMAND	
No. 3 Group	
Croft, Yorkshire	7 November 1942 to 31 December 1942
to No. 6 (RCAF) Group	
No. 6 (RCAF) Group	
Croft, Yorkshire	1 January 1943 to 4 May 1943
No. 63 (RCAF) Base	
Leeming, Yorkshire	5 May 1943 to 29 August 1945
to No. 1 Group	
No. 1 Group	
No. 63 (RCAF) Base	
Leeming, Yorkshire	30 August 1945 to 1 June 1946
Disbanded at Leeming	1 June 1946

REPRESENTATIVE AIRCRAFT TYPES

AIRCRAFT TYPE	ENGINE TYPE	DATES OF SERVICE
Vickers Wellington B.Mk.III Z1604 ZL*P	Bristol Hercules	November 1942 to March 1943
Vickers Wellington B.Mk.X	Bristol Hercules	February 1943 to May 1943
Handley Page H.P.63 Halifax B.Met Mk.V DK186 ZL*L "London's Revenge"	Rolls Royce Merlin	May 1943 to December 1943
H.P. Halifax B/A Met Mk.V LK900 ZL*D	Rolls Royce Merlin	October 1943 to January 1944
Handley Page H.P.61 Halifax B.Mk.III LW130 ZL*U	Bristol Hercules	January 1944 to March 1945
Avro Lancaster B.Mk.I PA271 ZL*W	Rolls Royce Merlin	March 1945 to May 1946
Avro Lancaster B.Mk.III RE160 ZL*G	Packard Merlin	March 1945 to May 1946

Handley Page Halifax

NO. 428 (BOMBER) "GHOST" SQUADRON RCAF
UNIT CODE   NA

No. 428(B) "Ghost" Squadron was formed at Dalton, Yorkshire on 7 November 1942, operating in No. 3 Group. The unit was equipped with Vickers Wellington B.Mk.III medium bombers. On 1 January 1943 the squadron was transferred to the newly formed No. 6 (RCAF) Group. Its first operational mission was flown on 26 January 1943 when six aircraft set out to bomb U-boat pens at Lorient, France. In April 1943 the unit converted to Vickers Wellington B.Mk.X aircraft.

In June 1943 the squadron relocated to Middleton St. George, Durham, where it converted to the Handley Page HP.63 Halifax B.Mk.V heavy bomber. In September, Handley Page HP.59 Halifax B.Mk.II aircraft were added to the operational inventory. In January conversion began to Handley Page HP.61 Halifax B.Mk.III aircraft, but this was suspended because of the decision to convert to the Avro Lancaster. The last operational mission flown in Halifax aircraft was in June 1944.

That month, the unit converted to the Canadian built Avro Lancaster B.Mk.X aircraft. The unit's final operational mission was flown on 25 April 1945 when fifteen aircraft bombed gun positions on the island of Wangerooge. At the end of hostilities the squadron flew its aircraft back to Canada.

Unit crest of No. 428 (Bomber) "Ghost" Squadron RCAF

Based at Yarmouth, Nova Scotia in 661 Heavy Bomber Wing (RCAF), Eastern Air Command, the squadron began training for participation in the invasion of Japan with RAF Tiger Force. The conversion to Canadian built Avro Lincoln B.Mk.XV aircraft was curtailed when the war with Japan ended. The squadron was disbanded at Yarmouth on 5 September 1945.

Pilot F/O Bill Blake and rear gunner W/O Jim Houston of 428(B) "Ghost" Sqn RCAF are pictured with their Handley Page HP.59 Halifax B.Mk.II at Dalton, Yorkshire on 17 March 1944. Both had recently been awarded the Distinguished Flying Cross for fighting off eleven attacks by German night fighters on a raid to Leipzig, Germany.
(CF PL-43473)

CHAPTER FOUR

HIGHER FORMATIONS AND BASES

BASE OF OPERATIONS	DURATION
BOMBER COMMAND	
No. 3 Group Dalton, Yorkshire to No. 6 (RCAF) Group	7 November 1942 to 31 December 1942
No. 6 (RCAF) Group Dalton, Yorkshire No. 64 (RCAF) Base Middleton St. George, Durham *en route to Canada*	1 January 1943 to 3 June 1943 4 June 1943 to 30 May191945 *31 May 1945 to 2 June 1945*
TIGER FORCE RCAF Eastern Air Command No. 6 (RCAF) Group No. 661 (RCAF) Heavy Bomber Wing Yarmouth, Nova Scotia Disbanded at Yarmouth	 25 July 1945 to 5 September 1945 5 September 1945

REPRESENTATIVE AIRCRAFT TYPES

AIRCRAFT TYPE	ENGINE TYPE	DATES OF SERVICE
Vickers Wellington B.Mk.III 　X3541　NA*N	Bristol Hercules	November 1942 to April 1943
Vickers Wellington B.Mk.X 　HE158　NA*L	Bristol Hercules	April 1943 to June 1943
Handley Page H.P.63 Halifax B.Mk.V 　W1271　NA*P	Rolls Royce Merlin	June 1943 to January 1944
Handley Page H.P.63 Halifax B/Met.Mk.V 　DK196　NA*Z	Rolls Royce Merlin	July 1943 to January 1944
Handley Page H.P.63 Halifax B/A.Met.Mk.V 　LK906　NA*D	Rolls Royce Merlin	September 1943 to November 1943
Handley Page H.P.59 Halifax B/GR Mk.II 　HR857　NA*K	Rolls Royce Merlin	September 1943 to June 1944
Avro Lancaster B.Mk.X 　KB760　NA*P	Packard Merlin	June 1944 to September 1945

Handley Page Halifax

NO. 429 (BOMBER) "BUFFALO" SQUADRON RCAF
UNIT CODE   AL

Formed at East Moor, Yorkshire as No. 429(B) "Buffalo" Squadron on 7 November, 1942, the unit operated in No. 3 Group, and was equipped with Vickers Wellington B.Mk.III medium bombers. On 1 January 1943 the squadron was transferred to the newly formed No. 6 (RCAF) Group and added Vickers Wellington B.Mk.X to its operational inventory. The squadron moved to Leeming, Yorkshire on 13 August 1943, where it was based until the end of the war.

At that time the squadron converted to the Handley Page HP.59 Halifax B.Mk.II heavy bomber. A gradual conversion to the HP.63 Halifax B.Mk.V, which began in November 1943, was suspended, and the Handley Page HP.61 Halifax was taken on strength beginning in January 1944. Halifax B.Mk.III, serial number LV993 coded AL*M was flown on ninety-seven operational missions with 429(B) Squadron before it was turned over to 1664 HCU. Its last mission was flown on 14 March 1945. This was the highest number of operations flown by any No. 6 (RCAF) Group aircraft. The last Halifax operational mission took place on 15 March 1945 against the Castrop-Rauxel oil refinery.

Conversion to the Avro Lancaster B.Mk.I and B.Mk.III began in March 1945. Operations continued with No. 6 Group until August, when the squadron transferred to No. 1 Group, RAF Bomber Command Strike Force, and began ferrying Allied prisoners of war and troops back to England from Italy. The unit was disbanded at Leeming on 1 June 1946.

Unit crest of No. 429 (Bomber) "Buffalo" Squadron RCAF

Members of 429(B) Sqn RCAF assembled in front of a Handley Page HP.63 Halifax B.Mk.V Series IA with the bison head, presented to the squadron by Canadian National Railways, 14 January 1944.

(CF PL-26849)

CHAPTER FOUR

HIGHER FORMATIONS AND BASES

BASE OF OPERATIONS	DURATION
BOMBER COMMAND	
No. 3 Group	
East Moor, Yorkshire	7 November 1942 to 31 March 1943
to No. 6 (RCAF) Group	
No. 6 (RCAF) Group	
No. 62 (RCAF) Base	
East Moor, Yorkshire	1 April 1943 to 12 August 1943
No. 63 (RCAF) Base	
Leeming, Yorkshire	13 August 1943 to 29 August 1945
to No. 1 Group	
No. 1 Group	
No. 63 (RCAF) Base	
Leeming, Yorkshire	30 August 1945 to 1 June 1946
Disbanded at Leeming	1 June 1946

REPRESENTATIVE AIRCRAFT TYPES

AIRCRAFT TYPE	ENGINE TYPE	DATES OF SERVICE
Vickers Wellington B.Mk.III BK163 AL*H	Bristol Hercules	November 1942 to August 1943
Vickers Wellington B.Mk.X HZ260 AL*K	Bristol Hercules	January 1943 to August 1943
Handley Page H.P.59 Halifax B.Mk.II LW285 AL*X	Rolls Royce Merlin	August 1943 to January 1944
Handley Page H.P.59 Halifax B/GR.Mk.II JD164 AL*K	Rolls Royce Merlin	August 1943 to January 1944
Handley Page H.P.63 Halifax B/Met Mk.V LK662 AL*Q	Rolls Royce Merlin	November 1943 to March 1944
Handley Page H.P.63 Halifax B/A.Met .Mk.V	Rolls Royce Merlin	November 1943 to March 1944
Handley Page H.P.61 Halifax B.Mk.III LV993 AL*M	Bristol Hercules	January 1944 to March 1945
Handley Page H.P.61 Halifax B/A.Mk.III LW688 AL*J	Bristol Hercules	March 1944 to May 1944
Avro Lancaster B.Mk.I PA226 AL*H	Rolls Royce Merlin	March 1945 to May 1946
Avro Lancaster B.Mk.III ME543 AL*B	Packard Merlin	March 1945 to May 1946

NO. 431 (BOMBER) "IROQUOIS" SQUADRON RCAF
UNIT CODE   SE

Formed at Burn, Yorkshire as No. 431(B) "Iroquois" Squadron on 11 November 1942, operating in No. 4 Group, the unit was equipped with Vickers Wellington B.Mk.X medium bombers. On 1 January 1943 the squadron was transferred to the newly formed No. 6 (RCAF) Group. The first operational mission was flown on 2 March 1943 when seven aircraft were despatched to lay mines in the area of the Frisian Islands.

On 15 July 1943 the unit moved to Tholthorpe converting at the same time to the Handley Page HP.63 Halifax B.Mk.V heavy bomber. On 10 December 1943 the squadron moved to Croft, remaining on that base to the end of the war. Conversion to Handley Page HP.61 Halifax B.Mk.III began in March 1944, although the B.Mk.V aircraft continued in service until July 1944. The last Halifax operational mission was flown on 25 October 1944.

Conversion to the Canadian built Avro Lancaster B.Mk.X began in October 1944, and these aircraft were flown until the end of the war. The last operational mission was flown on 15 April 1945, when aircraft were despatched to bomb gun positions on the island of Wangerooge. The squadron flew their Lancaster aircraft back to Canada beginning on 7 June 1945.

Based at Dartmouth, Nova Scotia, the squadron was attached to No. 662 Heavy Bomber Wing, Eastern Air Command to begin training for participation in the invasion of Japan as part of Tiger Force. The anticipated conversion to Canadian built Avro Lincoln B.Mk.XV bombers was curtailed when the war with Japan ended. The squadron was disbanded at Dartmouth on 5 September 1945.

Unit crest of No. 431 (Bomber) "Iroquois" Squadron RCAF

Handley Page HP.61 Halifax B.Mk.III heavy bombers of 431(B) "Iroquois " Sqn RCAF being prepared for a mission at Croft, Yorkshire, 22 June 1944. *(CF PL-40972)*

CHAPTER FOUR

HIGHER FORMATIONS AND BASES

BASE OF OPERATIONS	DURATION
BOMBER COMMAND	
No. 3 Group	
Burn, Yorkshire	11 November 1942 to 14 July 1943
to No. 6 (RCAF) Group	
No. 6 (RCAF) Group	
Tholthorpe, Yorkshire	15 July 1943 to 9 December 1943
No. 64 (RCAF) Base	
Croft, Yorkshire	10 December 1943 to 6 June 1945
en route to Canada	*7 June 1945 to 11 June 1945*
TIGER FORCE	
RCAF Eastern Air Command	
No. 6 (RCAF) Group	
No. 662 (RCAF) Heavy Bomber Wing	
Dartmouth, Nova Scotia	12 June 1945 to 5 September 1945
Disbanded at Dartmouth	5 September 1945

REPRESENTATIVE AIRCRAFT TYPES

AIRCRAFT TYPE	ENGINE TYPE	DATES OF SERVICE
Vickers Wellington B.Mk.X HE183 SE*J	Bristol Hercules	December 1942 to July 1943
Handley Page H.P.63 Halifax B/Met. Mk.V LK640 SE*Q "Queenie"	Rolls Royce Merlin	July 1943 to April 1944
Handley Page H.P.63 Halifax B/A.Met.Mk.V LK991 SE*U	Rolls Royce Merlin	September 1943 to February 1944
Handley Page H.P.61 Halifax B.Mk.III MZ600 SE*N MZ629 SE*B	Bristol Hercules	March 1944 to October 1944

Handley Page H.P.63 Halifax B/A.Mk.V Series IA
    LK640   SE*Q
    LW412   SE*F
	Rolls Royce Merlin	June 1944 to July 1944

Avro Lancaster B.Mk.X	Packard Merlin	October 1944 to September 1945

Handley Page Halifax

NO. 432 (BOMBER) "LEASIDE" SQUADRON RCAF
UNIT CODE   QO

Formed at Skipton-on-Swale as No. 432(B) "Leaside" Squadron on 1 May 1943, the unit operated in No. 6 (RCAF) Group, and was equipped with Vickers Wellington B.Mk.X medium bombers. The first operational mission was flown on 23 May 1943 when ten aircraft were despatched to bomb Dortmund, Germany. In September 1943 the unit moved to East Moor, and converted to Avro Lancaster B.Mk.II aircraft in October 1943.

In February 1944 the unit converted to Handley Page HP.61 Halifax B.Mk.III aircraft, and to the Handley Page HP.61 Halifax B.Mk.VII in June 1944. These were operated until the squadron disbanded on 15 May 1945. The last operational mission was flown on 25 April 1945 against gun emplacements on the island of Wangerooge.

Unit crest of No. 432 (Bomber) "Leaside" Squadron RCAF

A Handley Page HP.61 Halifax B.Mk.III of 432(B) "Leaside" Sqn RCAF is being armed for an operational mission at its base on 26 May 1944. In the foreground are 500-pound high explosive bombs ready to be loaded into the aircraft bomb bays. *(CF PL-29794)*

96

CHAPTER FOUR

HIGHER FORMATIONS AND BASES

BASE OF OPERATIONS	DURATION
BOMBER COMMAND	
No. 6 (RCAF) Group	
Skipton-on-Swale, Yorkshire	1 May 1943 to 18 September 1943
No. 62 (RCAF) Base	
East Moor, Yorkshire	19 September 1943 to 15 May 1945
Disbanded at East Moor	15 May 1945

REPRESENTATIVE AIRCRAFT TYPES

AIRCRAFT TYPE	ENGINE TYPE	DATES OF SERVICE
Vickers Wellington B.Mk.X HE222 QO*E	Bristol Hercules	May 1943 to October 1943
Avro Lancaster B.Mk.II DS848 QO*R	Bristol Hercules	October 1943 to February 1944
Handley Page H.P.61 Halifax B.Mk.III LK764 QO*F	Bristol Hercules	February 1944 to July 1944
Handley Page Halifax B/A.Mk.III LW583 QO*L	Bristol Hercules	cFebruary 1944 to June 1944
Handley Page H.P.61 Halifax B.Mk.VII NP707 QO*W "Willie the Wolf"	Bristol Hercules	July 1944 to May 1945

NO. 433 (BOMBER) "PORCUPINE" SQUADRON RCAF
UNIT CODE   BM

No. 433(B) Squadron was formed on 25 September 1943 at Skipton-on-Swale, Yorkshire, and operated in No. 6 (RCAF) Group. The unit was equipped with Handley Page HP.61 Halifax B.Mk.III aircraft, which first arrived at the squadron on 3 November. The first operational mission was flown on 2 January 1944, when four aircraft were despatched to lay mines in the area of the Frisian Islands. The last operation flown with Halifax aircraft took place on 16 January 1945.

The squadron converted to Avro Lancaster B.Mk.I and B.Mk.III bombers in January 1945. The last operational mission was flown on 25 April 1945 when ten aircraft bombed gun positions on the island of Wangerooge. After the end of hostilities, the squadron remained in England.

In August 1945 the unit was transferred to No. 1 Group in RAF Bomber Command Strike Force, engaged in ferrying prisoners of war back to England. The squadron disbanded at Skipton-on-Swale on 15 October 1945.

Unit crest of No. 433 (Bomber) "Porcupine" Squadron RCAF

Handley Page HP.61 Halifax B.Mk.III of 433(B) "Porcupine" Sqn RCAF at its dispersal, being prepared for an operational mission in the autumn of 1944. *(CF PL-43472)*

CHAPTER FOUR

HIGHER FORMATIONS AND BASES

BASE OF OPERATIONS	DURATION
BOMBER COMMAND	
No. 6 (RCAF) Group	
No. 63 (RCAF) Base	
Skipton-on-Swale, Yorkshire	25 September 1943 to 30 August 1945
to No. 1 Group	
No. 1 Group	
Skipton-on-Swale	31 August 1945 to 15 October 1945
Disbanded at Skipton-on-Swale	5 October 1945

REPRESENTATIVE AIRCRAFT TYPES

AIRCRAFT TYPE	ENGINE TYPE	DATES OF SERVICE
Handley Page H.P.61 Halifax B.Mk.III MZ857 BM*N "No Nuttins"	Bristol Hercules	November 1943 to January 1945
H.P.61 Halifax B/GR.Mk.III HX*230 BM*P	Bristol Hercules	December 1943 to August 1944
Avro Lancaster B.Mk.I NG441 BM*L "Little Lulu Mk.II"	Rolls Royce Merlin	January 1945 to October 1945
Avro Lancaster B.Mk.III SW273 BM*V	Packard Merlin	January 1945 to October 1945

Handley Page Halifax

NO. 434 (BOMBER) "BLUENOSE" SQUADRON RCAF
UNIT CODE WL

No. 434(B) Squadron was formed on 13 June 1943 at Tholthorpe, Yorkshire. The squadron operated in No. 6 (RCAF) Group, equipped with Handley Page HP.63 Halifax B.Mk.V heavy bombers. The first bombing mission was flown on 12 August when ten aircraft carried out a raid on Milan, Italy. The squadron moved to Croft, Yorkshire on 11 December, 1943. The unit began conversion to Handley Page HP.61 Halifax B.Mk.III in May 1944. The last operational mission with Halifax aircraft was flown on 21 December 1944 against Cologne, Germany.

Beginning in December 1944 the squadron was equipped with Canadian built Avro Lancaster B.Mk.X aircraft. The Lancasters were flown back to Canada in June 1945. Based at Dartmouth, Nova Scotia, the squadron was attached to No. 662 (RCAF) Heavy Bomber Wing, Eastern Air Command to began training for participation in the invasion of Japan as part of RAF Tiger Force. The anticipated conversion to Canadian built Avro Lincoln B.Mk.XV bombers was curtailed when the war with Japan ended. The squadron was disbanded on 5 September 1945 at Dartmouth.

Unit crest of No. 434 (Bomber) "Bluenose" Squadron RCAF

A Handley Page HP.63 Halifax B.Mk.V Series I (Special) of 434(B) "Bluenose" Squadron RCAF is refuelled in preparation for a bombing mission on 21 October 1943. *(CF PL-22138)*

100

CHAPTER FOUR

HIGHER FORMATIONS AND BASES

BASE OF OPERATIONS	DURATION
BOMBER COMMAND	
No. 6 (RCAF) Group	
Tholthorpe, Yorkshire	13 June 1943 to 10 December 1943
No. 64 (RCAF) Base	
Croft, Yorkshire	11 December 1943 to 9 June 1945
en route to Canada	*10 June 1945 to 14 June 1945*
RAF TIGER FORCE	
RCAF Eastern Air Command	
No. 6 (RCAF) Group	
No. 662 (RCAF) Heavy Bomber Wing	
Dartmouth, Nova Scotia	15 June 1945 to 5 September 1945
Disbanded at Dartmouth	5 September 1945

REPRESENTATIVE AIRCRAFT TYPES

AIRCRAFT TYPE	ENGINE TYPE	DATES OF SERVICE
Handley Page H.P.63 Halifax B.Mk.V EB217 WL*A	Rolls Royce Merlin	June 1943 to May 1944
Handley Page H.P.63 Halifax B/Met.Mk.V DK250 WL*W	Rolls Royce Merlin	August 1943 to March 1944
Handley Page H.P.63 Halifax B/A.Met.Mk.V LK893 WL*U	Rolls Royce Merlin	August 1943 to April 1944
Handley Page H.P.61 Halifax B.Mk.III LK799 WL*E	Bristol Hercules	June 1944 to October 1944
Handley Page Halifax B/A.Mk.III NA497 WL*C "Pubwash II"	Bristol Hercules	May 1944 to December 1944
Avro Lancaster B.Mk.X KB880 WL*L "Lollipop"	Packard Merlin	December 1944 to September 1945
Avro Lancaster B.Mk.I NG343 WL*U	Rolls Royce Merlin	February 1945 to March 1945
Avro Lancaster B.Mk.III PA228 WL*X	Packard Merlin	February 1945 to March 1945

Handley Page Halifax

HEAVY CONVERSION UNITS

At the Heavy Conversion Units, crews that had graduated from the Operational Training Units were instructed on the operation of four-engine heavy bomber aircraft which they would fly with the operational squadrons. The crews also learned survival, escape, and evasion tactics.

Training at the Heavy Conversion Units was hazardous. There were a multitude of accidents, ranging from mid air collisions to crashes on takeoff and landing. The aircraft used by training units had usually survived operational tours or had been overhauled and were deemed unsuitable to continue in operational use.

1659 HEAVY CONVERSION UNIT
Unit Code   RV/FD

Formed at Leeming, Yorkshire as No. 1659 HCU on 6 October 1942, the unit combined the Halifax Conversion Flights previously attached to both 405(B) "Vancouver" Squadron, RCAF, and 408(B) "Moose" Squadron RCAF. The unit moved to Topcliffe on 14 March 1943. On 20 November 1944 the unit was tasked with conversion of crews to the Canadian built Avro Lancaster B.Mk.X aircraft. The unit was disbanded on 10 September 1945.

REPRESENTATIVE AIRCRAFT TYPES

AIRCRAFT TYPE	ENGINE TYPE	DATES OF SERVICE
Handley Page H.P.57 Halifax B.Mk.I L9524 FD*D	Rolls Royce Merlin	October 1942 to March 1943
Handley Page H.P.59 Halifax B.Mk.II R9448 *L	Rolls Royce Merlin	December 1942 to May 1944
Handley Page H.P.59 Halifax B/GR.Mk.II HR801 RV*D	Rolls Royce Merlin	June 1943 to August 1944
Handley Page H.P.63 Halifax B.Mk.V LK864	Rolls Royce Merlin	April 1943 to October 1944
Handley Page H.P.63 Halifax B/Met.Mk.V LK703	Rolls Royce Merlin	August 1943 to June 1944
Handley Page H.P.63 Halifax B/A.Met.Mk.V LL542	Rolls Royce Merlin	August 1943 to October 1944
Handley Page H.P.61 Halifax B.Mk.III MZ505 RV*J HX268 FD*V	Bristol Hercules	November 1944 to May 1945
Handley Page Halifax B/A.Mk.III LW552	Bristol Hercules	January 1945 to February 1945
Avro Lancaster B.MK.X	Packard Merlin	November 1944 to September 1945

CHAPTER FOUR

Handley Page Halifax B.Mk.II Series IA aircraft of 1659 HCU at Croft, Yorkshire on 1 September 1943. The engines are being serviced and bombs loaded for a sortie on 22 November 1943. *(CF PL-26144)*

Handley Page Halifax

1664 HEAVY CONVERSION UNIT
Unit Code   ZU / DH (from 1944)

No. 1664 HCU was formed on 10 May 1943 at Croft, Yorkshire with Handley Page HP.63 Halifax B.Mk.V, and later HP.59 Halifax B.Mk.II aircraft. The unit was moved to Dishforth on 7 December, 1943 and nicknamed "Caribou" on 21 June, 1944. Conversion to Handley Page Halifax B.Mk.III began in December 1944. The unit was disbanded at Dishforth on 6 April 1945.

REPRESENTATIVE AIRCRAFT TYPES

AIRCRAFT TYPE	ENGINE TYPE	DATES OF SERVICE

Handley Page H.P.59 Halifax B/GR.Mk.II
    BB201   ZU*E
    JN968   DH*K
	Rolls Royce Merlin	August 1943 to October 1944

Handley Page H.P.63 Halifax B.Mk.V
    EB198   ZU*O
    DG280   DH*M
	Rolls Royce Merlin	May 1943 to April 1945

Handley Page H.P.63 Halifax B/Met.Mk.V DK248 ZU*C	Rolls Royce Merlin	November 1943 to November 1944

Handley Page H.P.63 Halifax B/A.Met.Mk.V
    LL137   DH*S
    LL194   ZU*R
	Rolls Royce Merlin	March 1944 to December 1944

Handley Page H.P.61 Halifax B.Mk.III
    MX587   ZU*G
    LK661   DH*N
	Bristol Hercules	December 1943 to Mach 1945

Handley Page H.P.61 Halifax B/A.Met.Mk.III NA498	Bristol Hercules	February 1945

A Handley Page HP.63 Halifax B.MK.V Series IA aircraft of 1664 HCU at Dishforth, Yorkshire on 26 June 1944. The aircraft is undergoing servicing at its dispersal. *(CF PL-30479)*

CHAPTER FOUR

1666 HEAVY CONVERSION UNIT
Unit Code   ND / QY

The unit was formed as No. 1666 HCU on 15 May 1943 at Dalton, Yorkshire with a complement of Handley Page HP.59 Halifax B.Mk.II and HP.63 Halifax B.Mk.V aircraft. The unit moved to Wombleton, Yorkshire on 21 October 1943, sharing the base with 1679 HCU which was a Lancaster conversion unit. The unit was nicknamed "Mohawk" on 21 June 1944.

When 1679 HCU was disbanded on 28 January 1944, the Lancaster conversion tasking was turned over to 1666 HCU, until the remaining Lancasters were transferred out in April 1944. At that time, additional HP.59 Halifax B.Mk.II and HP.63 Halifax B.Mk.V aircraft were taken on strength, and a three-flight system was implemented.

On 3 November 1944 Handley Page HP.61 Halifax B.Mk.III were received. However, the allocation was cancelled in favour of the Avro Lancaster on 20 November 1944. The first course on those aircraft began on 27 December 1944. The last Halifax B.Mk.V and B.Mk.III courses were completed by the end of January 1945, with the last Halifax aircraft departing in March. The unit was disbanded at Wombleton on 3 August 1945.

A Handley Page HP.59 Halifax B.Mk.II Series I (Special) of 1666 HCU is wheeled out of the maintenance hangar after overhaul at Wombleton, Yorkshire, on 28 June 1944. *(CF PL-30843)*

Handley Page Halifax

REPRESENTATIVE AIRCRAFT TYPES

AIRCRAFT TYPE	ENGINE TYPE	DATES OF SERVICE
Handley Page H.P.59 Halifax B/GR.Mk.II DT689 ND*K JD212 QY*H	Rolls Royce Merlin	December 1943 to December 1944
Handley Page H.P.59 Halifax B.Mk.II W1046	Rolls Royce Merlin	December 1943 to January 1945
Handley Page H.P.63 Halifax B.Mk.V DG296	Rolls Royce Merlin	May 1943 to January 1945
Handley Page H.P.63 Halifax B/A.Met.Mk.V LL519 QY*M	Rolls Royce Merlin	November 1944
Handley Page H.P.61 Halifax B.Mk.III PB578 ND*R	Bristol Hercules	November 1944 to March 1945
Handley Page H.P.61 Halifax B/A.Mk.III LW575	Bristol Hercules	November 1944 to March 1945
Avro Lancaster B.Mk.X	Packard Merlin	November 1944

CHAPTER FIVE

Halifax Colour Schemes and Markings

Members of 424(B) "Tiger" Sqn RCAF pose on the nose of a Handley Page HP.61 Halifax B.Mk.III bomber adorned with "Bambi," from the Walt Disney movie of that name, 4 December 1944. *(CF PL-40884)*

Handley Page Halifax

RAF Standard Night Bomber Colour and Marking Scheme

This finish and marking scheme was applied to the Handley Page Halifax aircraft as delivered from the factory up to early 1942.

5.1A Original Standard Night Bomber Three Colour Variegated Colour Scheme and Markings.

The upper surfaces of the fuselage, wings, engine nacelles, and horizontal stabilizers were finished in a variegated pattern of Dark Green and Dark Earth. The lower fuselage, vertical stabilizers and rudders, and all undersides were finished in Night (Black). On the fuselage sides, the demarcation between the upper surface variegated pattern and the lower surface colour was a wavy line, which had originally extended further down the side of the fuselage.

The national markings consisted of 84-inch Type B roundels, Red and Blue only, on the upper surfaces of both wings. On both sides of the rear fuselage, 64-inch Type A1 roundels were applied. Unit Codes and Aircraft Letter, in Light Grey, were applied, with the Unit Code generally to the left of the roundel and with the Aircraft Letter to the right. A flash was applied to both the inner and outer surfaces of the vertical stabilizers. The aircraft serial number was applied to both sides of the rear fuselage ahead of the horizontal stabilizer in 8-inch black characters. On the upper surface of the port wing, the dinghy stowage panel had a dull red sealing strip around the outer edges. There were no underside markings.

These marking features are presented in the following illustration.

1. ROUNDEL, Type A1
2. ROUNDEL, Type B
3. Serial Number
4. UNIT CODE and AIRCRAFT LETTER
5. FIN FLASH

CHAPTER FIVE

5.1B Modified Standard Night Bomber Three-Colour Variegated Colour Scheme with Revised Markings

This finish and marking scheme was applied to the Handley Page Halifax aircraft as delivered from the factory after early 1942.

The upper surfaces of the fuselage, wings, engine nacelles, and horizontal stabilizers were finished in a variegated pattern of Dark Green and Dark Earth. The lower fuselage, vertical stabilizers and rudders, and all undersides were finished in Night (Black), which offered less resistance to the airflow. On the fuselage sides, the demarcation between the upper surface variegated pattern and the lower surface colour was a straight line.

The national markings consisted of 84-inch Type B roundels, Red and Blue only, on the upper surfaces of both wings. On both sides of the rear fuselage, 64-inch Type C1 roundels were applied. Unit Codes and Aircraft Letter, in Dull Red, were applied, with the Unit Code to the left of the roundel and with the Aircraft Letter to the right (with some exceptions). A flash was applied to both the inner and outer surface of the vertical stabilizers; red, white and blue reading from front to rear. The aircraft serial number was applied to both sides of the rear fuselage in Dull Red characters ahead of the horizontal stabilizer. On the upper surface of the port wing, the dinghy stowage panel had a dull red sealing strip around the outer edges. There were no underside markings

These marking features are presented in the following illustration.

1. ROUNDEL, Type C1
2. ROUNDEL, Type B,
3. Serial Number
4. UNIT CODE and AIRCRAFT LETTER
5. FIN FLASH

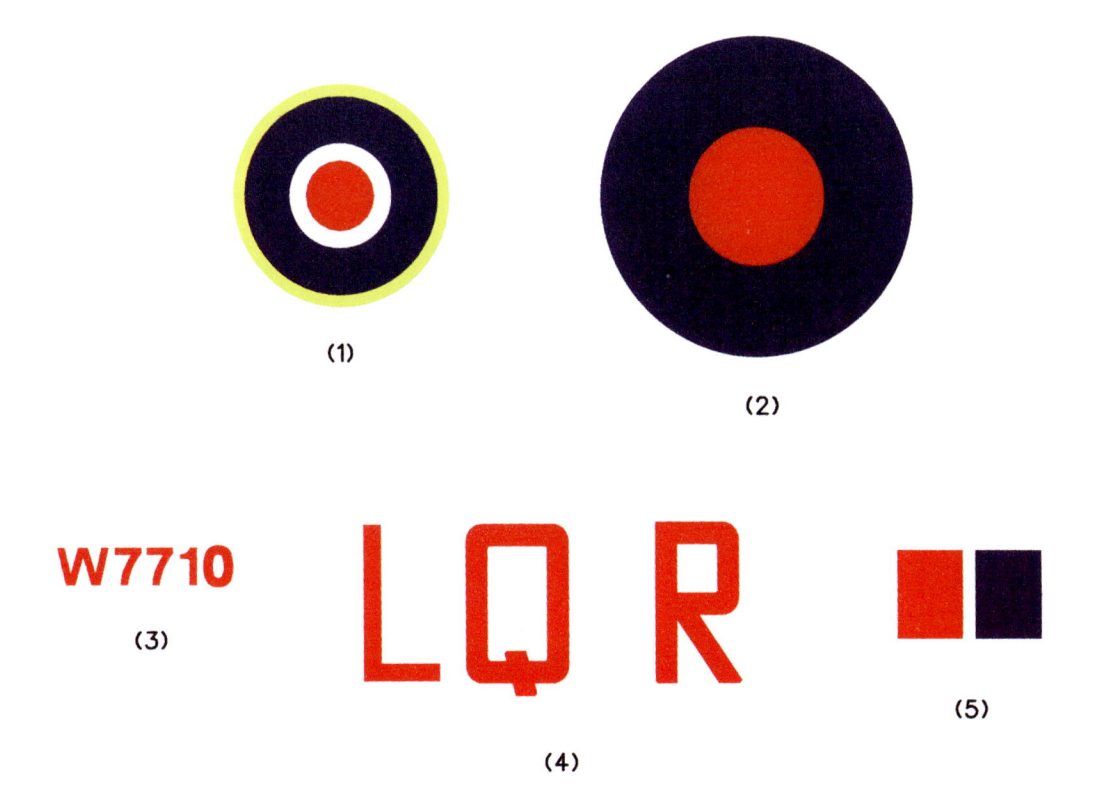

Handley Page Halifax

5.1C Nose Art

The Canadian bomber squadrons were noted for adorning their aircraft with "nose art." The aircraft letter usually was prominent in the name applied, but there were exceptions. The application of this decoration was the subject of two recent publications, *Metal Canvas: Canadians and World War II Aircraft Nose Art*, by Stephen Fochuk, and *RAF & RCAF Aircraft Nose Art In World War II*, by Clarence Simonsen. Both books have many fine examples of these decorations.

Members of 427(B) "Lion" Sqn RCAF carrying out maintenance on a Handley Page HP.63 Halifax B.Mk.V bomber. The 427 Squadron aircraft were adopted by Metro Goldwyn Mayer Studios, and named for studio stars. "London's Revenge" is decorated with the lion symbol of both the studio and the squadron, and is named for Lana Turner, whose name was carried on the nose of the aircraft. *(CF PL-26138)*

3. Handley Page HP.59 Halifax B.Mk.II Series I (Special), serial number DT689, coded VR*N of 419(B) "Moose" Squadron, RCAF, Middleton St. George, Durham, summer, 1943.

4. Handley Page HP.59 Halifax B.Mk.II Series IA, serial number JN969, coded NA*V of 428(B) "Ghost" Squadron RCAF, Middleton St. George, Durham, winter, 1943/44.

Illustrations by Stephen Otvos

1. Handley Page HP.57 Halifax B.Mk.I Series I, serial number L9524, coded RV*D of 1659 HCU RCAF, Topcliffe, Yorkshire, autumn, 1942.

2. Handley Page HP.59 Halifax B.Mk.II Series I, serial number W7710, coded LQ*R of 405(B) "Vancouver" Squadron RCAF, Topcliffe, Yorkshire, summer, 1942.

Illustrations by Stephen Otvos

5. Handley Page HP.63 Halifax B.Mk.V Series IA, serial number LL175, coded SE*A of 431(B) "Iroquois" Squadron RCAF, Tholthorpe, Yorkshire, summer, 1943.

6. Handley Page HP.61 Halifax B.Mk.III, serial number MZ954, coded KW*M of 425(B) "Alouette" Squadron RCAF, Tholthorpe, Yorkshire, autumn, 1944.

Illustrations by Stephen Otvos

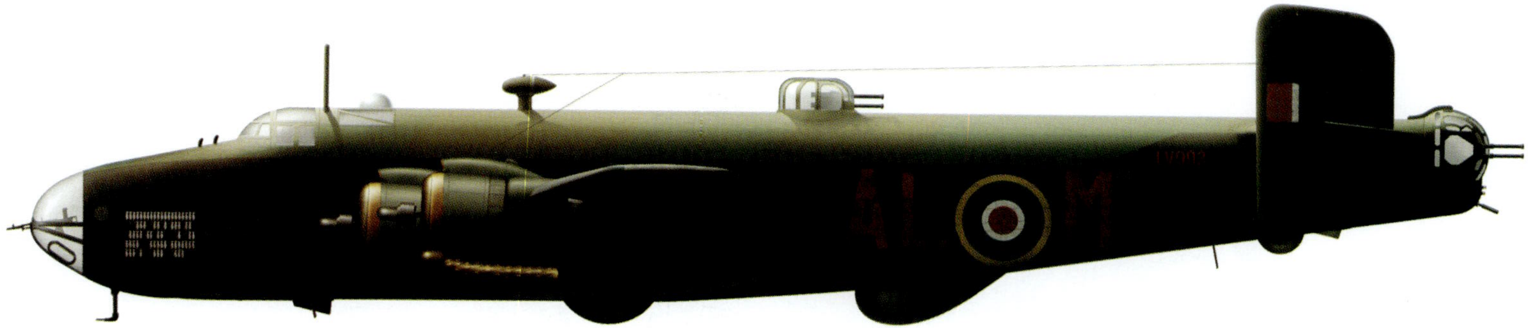

7. Handley Page HP.63 Halifax B.Mk.III, serial number LV993, coded AL*M of 429(B) "Bison" Squadron RCAF, Leeming, Yorkshire, spring, 1945.

8. Handley Page HP.63 Halifax B.Mk.VII, serial number NP761, coded EQ*A of 408(B) "Goose" Squadron RCAF, Linton on Ouse, Yorkshire, spring, 1945.

Illustrations by Stephen Otvos

CHAPTER SIX

Aircraft Armament Configurations and Ordnance

A tractor tows a train of bomb trolleys carrying a mixed load of armament, with a high explosive bomb in the foreground, to arm a Handley Page HP.59 Halifax B.Mk.II Series IA aircraft in preparation for a mission over enemy territory, 10 August 1943. *(CF PL-19506)*

Handley Page Halifax

On 7 July 1940 the two Halifax prototypes L7244 and L7245 were tested at Farnborough with mock-up turrets fitted. Subsequently Halifax B.Mk.I, L9485 was the aircraft used to test various armament configuration proposals. During air tests, severe vibration was experienced in the front of the fuselage caused by turbulence from the front turret. When the turret was turned to either side, the aircraft became difficult to control, and drag made returning the turret to the central position difficult. Although modifications were made to improve this situation, a yawing movement remained when the turret was turned.

Handley Page opted for the Boulton Paul turrets, introducing the HP.57 Halifax B.Mk.I into service with a power operated Type C in the nose position. The turret was powered by an electro-hydraulic system, with the power unit mounted in the front lower section of the chassis. Ammunition boxes holding 1,000 rounds for each gun were located either side of the gunner. The guns were mounted on their sides with cocking handles facing upwards, with the ammunition belts being drawn along rollers through openings in the gun trunnions from the ammunition boxes below. This turret received little use other than in daylight operations and was used to counter head-on attacks.

The Boulton Paul Type E turret was installed in the tail position. To the front of the gunner was the control table through which the operating stick protruded. Above the table was a panel facing the gunner with main motor switch, sight switch and oxygen supply socket. Arm rests were installed which lowered into position when the gunner was seated. Directly in front was the Mk.III reflector sight fixed to an arm connected to the gun arms. Below the gun was a 9mm armoured visor fixed on a frame which moved in elevation with the gunsight to provide frontal protection. Some of these were removed to provide a better field of view. All-round visibility was adequate, but the rear downward view was restricted by the guns and controls. This was partly remedied by the installation of two hydraulic rams which, when the guns were depressed, raised the seat maintaining the line of sight parallel to the gun barrels and giving a good view downwards.

In the beam positions, twin Vickers K guns were mounted. This feature was particular to the B.Mk.I only, protecting the aircraft from beam attack. It was suggested that the installation of a mid upper turret would improve the defensive armament.

On the HP.59 Halifax B.Mk.II Series I, a Boulton Paul Type C Mk.II mid upper turret was installed. This was the same as the nose turret, with a rear section added to accommodate the gunner. Although this installation afforded an improved defence in the beam and overhead zone, the increased drag caused by this turret installation reduced the top speed of the aircraft by 6 mph.

On night operations attacks from the front, beam and overhead sectors were infrequent. As a result, a proposal was put forward to remove the nose turret, and then the mid upper turrets. In the nose position a more streamlined fairing, called the "Z" nose fairing, was designed, and a blanking plate covered over the opening where the mid upper turret had been installed. This resulted in a weight saving of 1,450 lbs, an increase of 16 mph in top speed and a corresponding increase in range. This modification was applied to aircraft in service at the Rawcliffe Repair Depot, and to aircraft on the production line, resulting in the designation HP.59 Halifax B.Mk.II Series I (Special) and HP.63 Halifax B.Mk.V Series I (Special).

Concurrently, a streamlined fully transparent nose fairing was designed, and first made of metal for aerodynamic testing. On the test aircraft L9515, a Boulton Paul Type A. Mk.VIII four gun mid upper turret was installed. The new configuration resulted in the HP.59 Halifax B.Mk.II Series IA and HP.63 Halifax B.Mk.V Series IA. Even with the mid upper turret installed, a further 10 mph increase in top speed was realized. A single Vickers K .303 machine gun was installed in the new nose, as a "scare gun."

In the mid under position, a remote control twin .50 calibre Browning machine gun installation was designed, but proved to be prone to overheating. In its place the single .50 calibre Preston Green mounting was adopted. This was first installed after attacks from the underside by German night fighters with upward firing cannons. This gun mounting was initially installed in place of the H2S radar installation which was in short supply, and was later replaced by the radar.

In 1937 a new series of bombs was adopted by the RAF which were aerodynamically shaped with tail fins. In the early years of the Second World War Bomber Command relied primarily on the 250 lb. and 500 lb. General Purpose (GP) high explosive bombs. Bombs were classified by their charge-to-weight ratio, or the comparison of explosive to gross weight. The GP bombs had a charge-to-weight ratio of 30 to 35 percent.

Heavy bombs, aerial mines, and types of small bomb containers carried by Halifax bombers *(Andrew Tattersall)*

RAF 2,000 & 4,000 LB HC BOMBS

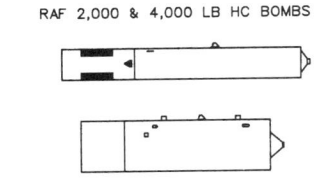

RAF 8,000 LB HC BOMB

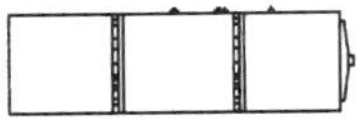

RAF 1,500 LB MINE AND 4 × 250 LB DEPTH CHARGES

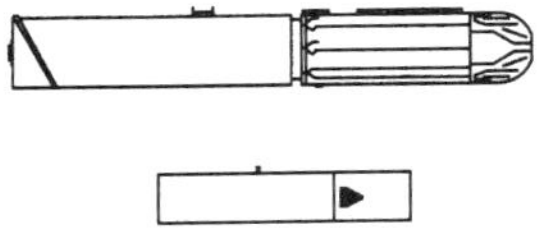

RAF SMALL BOMB CONTAINERS (SBC)

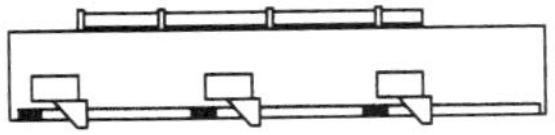

Various armament configurations in Halifax aircraft *(Andrew Tattersall)*

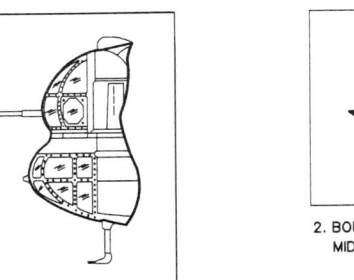

1. BOULTON PAUL TYPE C Mk V NOSE TURRET
2. BOULTON PAUL TYPE C Mk II MID UPPER FUSELAGE TURRET 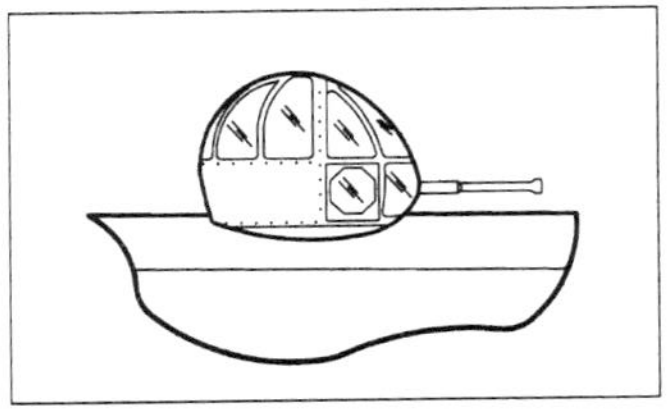
3. BOULTON PAUL TYPE E TAIL TURRET 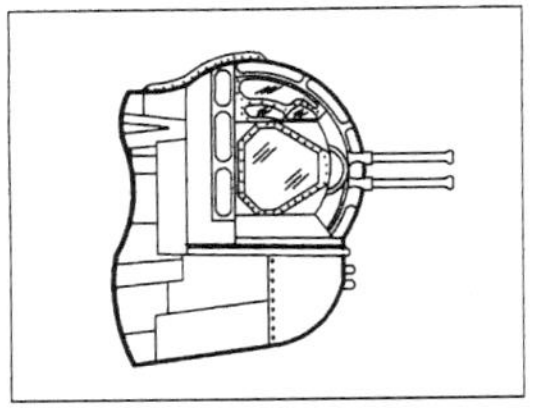
4. BOULTON PAUL TYPE A MID UPPER FUSELAGE TURRET 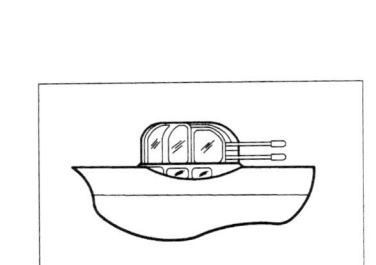
5. CLEAR PERSPEX NOSE FAIRING WITH .303 CALIBRE SCARE GUN 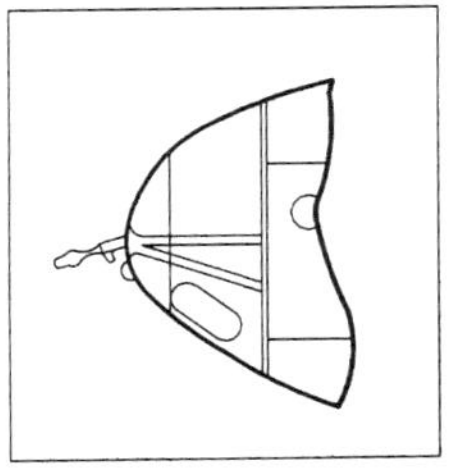
6. PRESTON GREEN UNDER FUSELAGE .50 CALIBRE DEFENCE MOUNTING

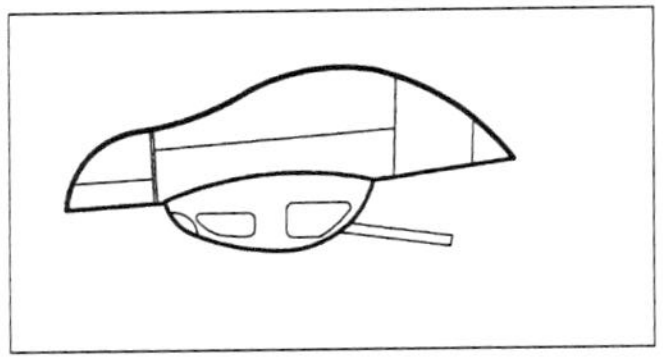

Handley Page Halifax

Development of new weapons with lighter weight casings led to the 500 lb., 750 lb., 1,000 lb, and 2,000 lb. Medium Capacity bombs, with a charge-to-weight ratio of 40 to 50 percent. Finally, the 4,000 lb.High Capacity (HC) "Cookie" greatly increased the offensive power of the RAF bomber force. These HC bombs had a charge-to-weight ratio of 75 to 80 percent. A further development was the 8,000 lb. HC Cookie (two 4,000 lb. Cookies bolted together), and the 12,000 lb. "Blockbuster." Small Bomb Containers (SBC) were loaded with incendiary devices, and were dropped in conjunction with high explosive bombs

The aerial mine was designed to withstand drops from aircraft flying at 200 mph at altitudes varying from 100 to 15,000 feet. The mines could be detonated by contact or by magnetic triggers. For antisubmarine operations, the 500 lb. depth charge was also carried.

The Halifax carried a wide variety of ordnance in its role as a heavy bomber and mine laying aircraft. The deep bomb bay extended the length of the fuselage centre section, and bomb cells were built into the inner sections of the wings between the main spars.

In Chapter 3, the various combinations of bomb loads are described for each variant of the aircraft. These include combinations of these bombs:

500 lb. bombs in fuselage bomb bay
500 lb. or 750 lb. bombs in each of six wing bomb bays
2,000 lb. bombs in fuselage bomb bay
1,000 lb. bombs in fuselage bomb bay
1,500 lb. aerial mines in fuselage bomb bay
4,000 lb. bombs in fuselage bomb bay
8,000 lb. bomb in fuselage bomb bay
Small bomb containers (SBC) in fuselage bomb bay with incendiaries

Viewed from the end of the line, the small bomb containers holding incendiaries are carried on the trailing bomb trolleys. A high explosive bomb is being prepared for loading under the aircraft, 10 August 1943. *(CF PL-19508)*

CHAPTER SEVEN

RCAF Memorial Museum Halifax Restoration

The salvaged components of Handley Page HP61 Halifax B.Mk.VII, serial number NA337, laid out in the restoration area of the RCAF Memorial Museum at CFB Trenton. *(CF TNC 96-65-1)*

Handley Page Halifax

Of 6178 Halifax bombers built during the Second World War, Halifax A.Mk. VII, serial number NA 337 is one of the three Halifax aircraft that remain. It was built by Rootes Securities at Speke. The aircraft was assigned to No. 644 Squadron RAF on 3 March 1945, and was coded 2P*X. The unit was based at Tarrant Rushton, Dorsetshire, operating in 38 Group on "Special Duties," assigned to the Airborne Forces.

When not engaged in airborne operations, the squadron aircraft were dropping agents and supplies to members of the Resistance groups in Europe. The supplies usually consisted of containers of weapons, ammunition and explosives. The containers, about 15 inches in diameter and 60 inches long, were carried in the bomb bays. One of these containers was recovered from the NA337 drop site in Norway and is on display at the Museum.

The first reference to NA337 in the 644 Squadron operations logs is on 24 March 1945, when the aircraft towed a General Aircraft "Hamilcar" glider, containing a Dodge truck and a 17-pounder gun, on the last great Airborne operation of the Second World War. Operation Varsity was the massive Airborne assault across the Rhine River into Germany. In the late evening of April 23 1945, just fifteen days before the end of the war in Europe, NA 337 took off from RAF Station Tarrant Rushton in England on a mission to drop supplies to the Norwegian Underground forces near Grue, Norway.

After a successful airdrop early the following morning, NA 337 started the return journey to England. Unfortunately its heading took it over an important and well-protected bridge, at the southern end of Lake Mjosa, where it came under accurate fire from a German anti-aircraft installation. A shell penetrated the starboard wing, igniting the fuel tank, and causing the two engines on that side to fail. Faced with no other alternative, the pilot, Flight Lieutenant A. Turnbullt was forced to ditch the aircraft in the lake. The landing was rough but successful, and all of the crew of six managed to escape before it sank in 770 feet of water.

Due to the cold temperatures, only one person, Flight Sergeant Thomas Weightman, the rear-gunner, managed to survive. Some six hours after the crash, he was found floating on an overturned dinghy. Later he was taken by the Germans to Oslo to await transport to a prison camp in Germany. Fortunately, the war ended before this could occur and he was repatriated to England.

For the next forty years NA 337 lay in the icy depths of Lake Mjosa, about two kilometres from shore. In 1982 it was located by two local Norwegian historians, Tore Marsoe and Rolf Liberg. Marsoe, as a boy of sixteen, had heard the aircraft crash in 1945, and had never lost interest in it. Together with Liberg he finally located the aircraft in 1982 using sonar equipment. However, with no official interest in the salvage of the aircraft, it remained in the depths of the lake.

In the intervening years, Karl Kjarsgaard, a Canadian Airlines pilot, heard mention of the downed Halifax. Together with Jeff Jeffery DFC, a second world war Halifax pilot, he came up with a plan to salvage and restore the aircraft. In 1994 the Halifax Aircraft Association was formed to provide impetus and seek funding and assistance for the project. Salvage rights were obtained from the Norwegian government, and the Canadian Forces agreed to provide a twelve-man engineering team to dismantle and preserve the aircraft as it was brought ashore.

On 5 September 1995, after a great deal of work by many dedicated individuals from Canada and Norway, and with Flt Sgt. Weightman in attendance, NA 337 finally broke the surface of Lake Mjosa again. Subsequently, it was taken by Canadian Forces CC130 Hercules aircraft from 8 Wing, Trenton to its final destination, the RCAF Memorial Museum, in August 1995.

The museum was founded by Captain Earl Hewison along with Major Ralph Patrick in 1983/84, and opened to the public in April 1984 in a temporary location, in the annex to the CFB Trenton Base Recreation Centre. The museum moved into the former indoor curling rink in 1993, and after refurbishment of the building, was opened to the public on 1 April 1994.

In 1996, the workshop area was constructed on the back of the main building, and Brigadier General (retd) Jeff Brace became Executive Director. He was succeeded by Lieutenant Colonel (retd) F. Chris Colton in 2002, under whose able and enthusiastic direction the museum continues to acquire and display historic aircraft and artifacts, and to build an historic archive of publications and documents.

After unloading at Trenton, work began on the long-term restoration of the Halifax. The aircraft was moved several times before coming to rest in the workshop area in the autumn of 1996. A dedicated team of volunteers under the direction of coordinator Lieutenant Colonel (retd) Bill Tytula has gradually

CHAPTER SEVEN

The new addition to the RCAF Memorial Museum at CFB Trenton designed to house the restored Halifax aircraft. *(RCAF Memorial Museum)*

restored the aircraft components and assemblies to near completion.

In May 2004 construction of the new addition to the museum which now houses the Halifax was started. The partially completed aircraft was moved into this facility on 16 October 2004. The new addition is scheduled for completion in 2007.

The location and recovery of yet another Halifax bomber provided components which became part of the restoration of the RCAF Memorial Museum aircraft. In 1984, Jay Hammond began to search out the location of Halifax B.Mk.III serial LW682 coded OW*M of 426(B) Squadron RCAF that had been flown by his uncle, Pilot Officer Wilbur Boyd Bentz. The aircraft had been bombing the railway yards at Louvain, Belgium on 12 May 1944 and crashed near Geraardsbergen (Grammont). Unfortunately the entire crew perished. When the German forces arrived at the site, they could only recover the remains of five airmen, four Canadian and one British. A German document written at the time the recovery unit left the site states that they could recover almost nothing from the plane because it had crashed in the boggy ground near the Dender river.

In 1985 Jay Hammond received a letter on a fisheries issue from Frank Mergen, a Luxembourg national who was studying pharmacy at the University of Louvain in Belgium. When Jay, a fisheries biologist with the province of British Columbia replied, knowing the bombing destination of his uncle's aircraft had been Louvain, he added an aside to that effect. Frank started making enquiries of Belgian citizens and eventually met eyewitnesses. He travelled to the site, and passed on all he learned to Jay. At last knowing the location of the aircraft, Jay then sought help in the recovery effort.

Karl Kjarsgaard had been working on restoring the Halifax Bomber at Trenton that had been recovered from Norway the year before. The request for Karl's help was met with an immediate affirmative. Karl approached the 426 Squadron Association to request their sponsorship of the project. After receiving confirmation of their support he developed a budget and began talking to officials at Canadian Heritage and Veterans Affairs Canada to obtain advice, funding support for a video of the recovery effort, and commitment to conduct a military funeral if remains of the airmen were found.

Handley Page Halifax

On 5 September Jay arrived in Geraardsbergen where he met with Karl and members of the Belgian Aviation History Association. At the site, the Canadians received all manner of support and assistance from a volunteer force of Belgians. The crash site was a bog, with the aircraft wreckage about 20 feet below the surface. Crews began pumping out the water on 30/31 August.

On 6 September the team successfully recovered the remains of the three missing airmen. Veterans Affairs Canada, the Canadian Forces and the 426 (Thunderbird) Squadron Association held a funeral service on 10 November at the Geraardsbergen (Gammont) Communal Cemetery for the three men. The next of kin of most of the eight crew members, members of the 426 Squadron Association, the Air Force Association of Canada, the RCAF POW Association, the Belgian Aviation History Association and representatives of the Belgian government were present.

The Canadians remained in Belgium cleaning and separating parts of the aircraft, returning home 15 September. Parts of the aircraft, notably the landing gear, were used to restore the Halifax NA337 recovered from Norway.

In the meantime, an effort has been launched to recover yet another Halifax aircraft. In 2005 a Canadian group will initiate an attempt to recover a Handley Page HP.61 Halifax B.Mk.III bomber from the north Atlantic seabed off Ireland The aircraft had served with a Halifax squadron of the RCAF during the latter part of the war.

In August 1945, while on patrol from its base in Scotland, Halifax serial number LW170 sprung a fuel leak and was forced to ditch in the sea about 130 miles northwest of Ireland. The aircraft remained afloat for seven hours. The crew took to life rafts and were rescued by a passing freighter.

Karl Kjarsgaard, who has participated in the recovery of both the RCAF Memorial Museum Halifax aircraft and the aircraft in Belgium is once again involved in this effort. The aircraft, if recovered, will be restored and displayed in the museum located in Nanton, Alberta.

CHAPTER EIGHT

Modelling the Handley Page Halifax

ABOVE AND AT RIGHT: Halifax B.Mk.II Series I, serial number W7710, coded LQ*R, named "The Ruhr Express," of 405(B) "Vancouver" Squadron RCAF. The markings were obtained from Skymodels Decal No. SKY 72-025. *(Model and photos by Bill Scobie)*

Handley Page Halifax

Since the aircraft was first introduced into service during the Second World War, the Handley Page Halifax has been a subject of interest to modellers. Only two plastic kits of the Halifax are known to have been available in 1:72 scale, while a new kit in 1:48 scale has been announced.

Models in 1:72 scale

MERLIN ENGINE VARIANTS

The Matchbox PK-604 Halifax kit offers the options to assemble a B.Mk.I Series I, a modified B.Mk.II Series I (without the Boulton Paul Type C mid upper turret), and a B/Gr.Mk.II Series Ia with the Boulton Paul Type A.Mk.VIII mid upper turret.

The Halifax B.Mk.I Series I, can be assembled from the Matchbox Kit No. PK-604 with no modifications.

Using the Matchbox kit and the Aeroclub Halifax "Z" nose conversion (C007), a Halifax B.Mk.II Series I Special, can be modelled. A B.Mk.V Series I Special, along with the Aeroclub "Z" nose conversion, also requires the replacement of the kit main landing gear with the Aeroclub (CV098) Dowty link undercarriage in white metal. The 4 blade Rotol propeller installed on some B.Mk.II or B.Mk.V aircraft is also available in white metal from Aeroclub (CP114), as is the standard 3 blade Rotol.propeller (CP115).

Using the Matchbox kit, the Halifax B.Mk.II Series IA or B.Mk.V Series IA can be modelled. The B.Mk.V Series IA again requires the replacement of the kit main landing gear with the Aeroclub CV098 Dowty link undercarriage in white metal.

HERCULES ENGINE VARIANTS

The Airfix Halifax kit No. 06008 assembles as a Halifax B.Mk.III with short wingspan and H2S radar installation. With minimal modification, the late B.Mk.III as well as a B.Mk.VI and B.Mk.VII can also be represented.

The Halifax B.Mk.III is modelled, with no change, using the Airfix kit. This kit features the short span wings of production aircraft and the H2S radome. Some early production aircraft had a ventral gun emplacement with a .50 cal. machine gun installed in place of the radar, while others had neither the gun nor the radar. Late production B.Mk.III aircraft had the long span wings, and both upper turret and H2S radar installation. The extended wingtips of the late model aircraft must be scratch built.

The Halifax B.Mk.VI can be modelled using the Airfix B.Mk.III kit. All B.Mk.VI aircraft featured the long span wings and had the H2S radar installed. The Halifax B.Mk.VII can also be modelled using the Airfix B.Mk.III kit. All B.Mk.VII aircraft featured the long span wings and had the H2S radar installed.

CHAPTER EIGHT

Handley Page HP.61 Halifax B.Mk.VII, serial number PN230, coded EQ*V and named "Vicky the Vicious Virgin" of 408(B) "Goose" Squadron RCAF, Linton-on-Ouse, Yorkshire, spring 1945.
(Model and photos by Bill Scobie)

Handley Page Halifax

The Matchbox Handley Page Halifax B.Mk.I/II kit No. PK-604 in 1:72 scale. *(Andrew Tattersall)*

The Airfix Handley Page Halifax B.Mk.III kit in 1:72 scale.

HANDLEY PAGE HALIFAX KITS IN 1:72 SCALE	KIT MANUFACTURER
Handley Page H.P.57 Halifax B.Mk.I Series I	Matchbox
Handley Page H.P.59 Halifax B.Mk.II Series I	Matchbox
Handley Page H.P.59 Halifax B.Mk.II Series IA	Matchbox

HANDLEY PAGE HALIFAX KIT IN 1:72 SCALE	KIT MANUFACTURER
Handley Page H.P.61 Halifax B.Mk.III, Short span wings	Airfix

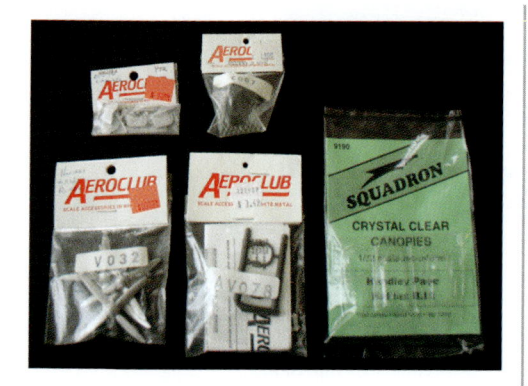

Some after-market items to improve the Halifax plastic kits. *(Andrew Tattersall)*

AFTER MARKET ITEMS

Item	Source
Halifax B.Mk.III Crystal Clear Canopy	Squadron 9190
Boulton Paul upper turret (B.Mk.II Series I)	Aeroclub CV078
"Z" nose conversion (B.Mk.II/V Series I Special)	Aeroclub C007
Dowty link undercarriage in white metal (B.Mk.V)	Aeroclub CV098
Delaney Gallay radiators (B.Mk.II/V)	Aeroclub CV034
Hercules Engine air intakes (B,Mk,III, VI, Vii)	Aeroclub CV036
Rotol 3 blade propellers in white metal	Aeroclub CP115
Rotol 4 blade propellers in white metal	Aeroclub CP114

126

CHAPTER EIGHT

The Belcher Bits BL2 heavy bomb set in 1:72 scale.
(Andrew Tattersall)

The Skymodels SKY 72-025 decal set in 1:72 scale.
(Andrew Tattersall)

AFTER MARKET ITEMS

RAF 2,000, 4,000, 8,000, 12,000 lb. HC Bombs	Belcher Bits BL2

DECALS

Handley Page Halifax	Skymodels SKY 72-025

Fonderie Miniature Halifax B.Mk.III kit in 1:48 scale.
(Fonderie Miniature)

The various Belcher Bits ordnance sets in 1:48 scale.
(Andrew Tattersall)

HANDLEY PAGE HALIFAX KITS IN 1:48 SCALE

	KIT MANUFACTURER
Handley Page Halifax B.Mk.III (for 2005 release)	Fonderie Miniature FN6042

AFTER MARKET ITEMS

RAF 2,000 & 4,000 lb. HC Bombs	Belcher Bits BB9
RAF 8,000 & 12,000 lb. HC Bombs	Belcher Bits BB10
RAF 1,500 lb. Mine & Four 250 lb. Depth Charges	Belcher Bits BB11
RAF Small Bomb Containers (SBC)	Belcher Bits BB17

127

BIBLIOGRAPHY

Barnes, C.H. *Handley Page Aircraft Since 1907*.London: Putnam & Co., 1987

Bowyer, M. and J. Rawlings. *Squadron Codes 1937-56*. Cambridge: Patrick Stevens Ltd, 1979

Goulding, James and Philip J.R. Moyes. *RAF Bomber Command and its Aircraft 1936-1940*. Shepperton, Surrey: Ian Allan Ltd, 1975

Griffin, J. *Canadian Military Aircraft: Serials and Photographs 1920-1968*.Ottawa: National Museum of Man, 1969

Halley, James J. *The Squadrons of the RAF & Commonwealth 1918-1988*. London: Air-Britain, 1980

Kostenuk, S. and J. Griffin. *RCAF Squadron Histories and Aircraft 1924-1968*. Toronto: Samuel Stevens Hakkert & Co., 1977

Moyes, Philip J.R. *The Handley Page Halifax B.III, VI, VII*. Profile Publications No.11, 1965

Thetford, Owen. *Aircraft of the Royal Air Force since 1918*. London: Putnam & Co., 1970

405 Squadron History. Winnipeg: Craig Kelman & Associates Ltd., 1986

408 Squadron History. Belleville: The Hangar Bookshelf, 1984

419 Squadron History. Glen Foulds. 419 Squadron, Cold Lake, 1991

424 Squadron History. Capt. Nora Bottomley. Belleville: The Hangar Bookshelf, 1985

426 Squadron History. Capt Ray Jacobson, 426 Squadron, Trenton, 1989

433 Squadron History. Capt J.S.R. Bastien. Belleville:The Hangar Bookshelf, 1985

434 Squadron History. Belleville: The Hangar Bookshelf, 1984

ADDITIONAL INFORMATION WAS FOUND IN THE FOLLOWING PERIODICALS:

"Bomber Command's Second String: The Halifax." *Air Enthusiast* No.15, 11-30

"Forties Favourites 1." *Aeroplane Monthly* (May 1987) 248-51, 269

Lintonian. "Halifax Part 1." *Air Pictorial* (December, 1963), 367-69

Lintonian. "Halifax Part 2." *Air Pictorial* (January, 1964), 11-13

Lintonian. "Halifax Part 3." *Air Pictorial* (February, 1964), 48-50

"Halifax Salvage." *Flypast* (November, 1996), 54-8

"Halifax Revelation." *Flypast* (September, 1998), 8

Merrick, K.A. "Handley Page Halifax." Aircraft Described No.177. *Aero Modeller* (December, 1968), 650-54

Sturtivant, Ray. "Aircraft in Detail: Handley Page Halifax." *Scale Aircraft Modelling* 10/8 (May 1988), 340-356

"Trenton and the Halifax." *Flypast* (November, 1998), 104-110